Manfred Hesse

Ring Enlargement
in Organic Chemistry

© VCH Verlagsgesellschaft mbH, D-6940 Weinheim (Federal Republic of Germany), 1991

Distribution:

VCH, P. O. Box 101161, D-6940 Weinheim (Federal Republic of Germany)

Switzerland: VCH, P. O. Box, CH-4020 Basel (Switzerland)

United Kingdom and Ireland: VCH (UK) Ltd., 8 Wellington Court, Cambridge CB1 1HZ (England)

USA and Canada: VCH, Suite 909, 220 East 23rd Street, New York, NY 10010-4606 (USA)

ISBN 3-527-28182-7 (VCH, Weinheim)   ISBN 0-89573-991-7 (VCH, New York)

Manfred Hesse

Ring Enlargement in Organic Chemistry

VCH   Weinheim · New York · Basel · Cambridge

Professor Dr. Manfred Hesse
Organisch-Chemisches Institut
der Universität Zürich
Winterthurer Straße 190
CH-8057 Zürich

Published jointly by
VCH Verlagsgesellschaft mbH, Weinheim (Federal Republic of Germany)
VCH Publishers, Inc., New York, NY (USA)

Editorial Director: Dr. Michael G. Weller
Production Manager: Claudia Grössl

Library of Congress Card No. applied for

British Library Cataloguing-in-Publication Data
Hesse, Manfred 1935–
Ring enlargement in organic chemistry.
1. Organic compounds. Synthesis
I. Title
547.2
ISBN 3-527-28182-7 Germany

CIP-Titelaufnahme der Deutschen Bibliothek
Hesse, Manfred:
Ring enlargement in organic chemistry / Manfred Hesse. –
Weinheim; New York; Basel; Cambridge: VCH, 1991
ISBN 3-527-28182-7 (Weinheim . . .)
ISBN 0-89573-991-7 (New York . . .)

Composition: Hagedornsatz, D-6806 Viernheim
Printing: Diesbach Medien, D-6940 Weinheim
Bookbinding: Großbuchbinderei J. Schäffer, D-6718 Grünstadt
Printed in the Federal Republic of Germany

For Barbara and Mickey

Preface

I have long been fascinated by the phenomenon of ring enlargement reactions. We had already in the late 1960s encountered this problem in studies aimed to clarify the structure of the spermidine alkaloids of the oncinotine and inandenine type. The ease with which a ring enlargement occurs, quite unprovoked, was baffling, and opened new perspectives. Since then many collaborators in my research team have sought with enthusiasm and persistence to develop these reactions in a methodical fashion and to harness them to the synthesis of natural products. When I was asked about a year ago whether I was finally ready to write a survey of the methodology of ring enlargement reactions, I readily agreed. A period of sabbatical leave linked to the task was equally tempting. With its help, so I thought, and free from the duties of teaching and administration, it would be an easy task to concentrate on a branch of science which seemed to me of the highest interest. I greatly looked forward to it – and accepted with the warmest gratitude the readiness of my colleagues in the Institute of Organic Chemistry to take over my work in the Institute, and so to provide the vital prerequisite of my scheme.

At first all went as we had hoped. I settled to concentrated study, provided with ample literature and good materials of work – in a quiet and peaceful cell, attended by my wife, who contrived to bring sympathy and understanding to an extraordinary degree to a branch of science wholly unknown to her, and to offer suggestions and improvements. Our sons, too, showed enthusiastic interest. But soon the grey light of everyday life crept into this idyll. The studies of my diploma and doctoral students still had to be corrected and examined; and though all were as considerate as possible – for which I would like once more heartily to thank my colleagues, diploma and doctoral students and postdoctoral fellows – I was drawn in to help solve problems in their work and into discussions with them. Furthermore, the material I had to digest proved to be far more copious than I had expected, and exceedingly difficult to master. In short, the relaxing scientific stroll in a lush, narrow valley grew more and more into a trek up an extremely steep and stony path, only to be conquered by calling out all my reserve.

To all those who shared in this enterprise I am more than grateful for their understanding while it was in the making. I must first thank my Secretary,

Mrs Martha Kalt, who photocopied the literature and processed my manuscript with tireless devotion. Mrs Esther Illi prepared the drawings in admirable fashion. I have to thank Professor Heinz Heimgartner for his valuable advice in the revision of the book, and Dr. Stephan Stanchev for much help in seeking out the literature. Dr. Volkan M. Kısakürek, Editor of Helvetica Chimica Acta, gave me unstinting aid in the production of the Index, for which I warmly thank him. Very grateful I am also to Prof. C. N. L. Brooke, Cambridge, for his kind help.

Last but not least, I owe warmest thanks to my friend James M. Bobbitt, Professor of Organic Chemistry in Storrs, Connecticut, who was most generously prepared to revise the English draft of the book and to make notable improvements.

Zürich, January 1991   M. H.

Contents

Drawing of the "ring enlarged" Tower Bridge by Jörg Kalt

I. Introduction

Chemists have been interested in macrocyclic compounds for more than sixty years. This era began in 1926 when Ruzicka published the structural elucidation of the musk components, civetone (Zibeton) and muscone [1]. Muscone was found to be 3-methylcyclopentadecanone (**I/1**). Soon afterwards, the presence of pentadecanolide (**I/2**) and 7-hexadecenolide (**I/3**) in the vegetable musk oils of *Angelica* roots (*Archangelica officinalis* Hoffm.) and ambrette seeds (*Abelmoschus moschatus* Moench), was discovered [2]. It was long before chemists tried to find synthetic routes to these and related macrocyclic cycloalkanones as well as to corresponding lactones. The cyclization reactions were studied carefully [3][1], and new techniques such as the dilution principle were developed. These materials were not only of scientific interest but of great commercial importance in the fragrance industry [4].

In the course of studying these reaction principles, the chemistry of medium and large ring compounds was investigated. This led to the discovery of the transannular reactions [5] which are a fascinating part of chemistry even today.

A second period of macrocyclic chemistry was signaled by the isolation of the first macrolide antibiotic from an *Actomyces* culture in 1950. Brockmann and Henkel [6][7] named it picromycin (Pikromycin) (**I/4**), because of its bitter taste. This antibiotic contains a 14-membered ring. Since then a large number of macrocyclic lactones, lactams and cycloalkane derivatives have been discovered. Some of these compounds have a considerable physiological importance for humans and animals. Because of these physiological properties it was necessary to prepare larger quantities of these macrocylic compounds by chemical syntheses [8].

The synthesis of macrocyclic compounds can be accomplished by ring forming or by ring enlargement processes. The starting materials for the ring enlargement approach are, of course, cyclic compounds themselves, presumably easier to prepare than the ultimate product.

An astonishing number of ways have been discovered to enlarge a given ring by a number of atoms. As will be shown in this review, the catalogue of the

---

1) Cyclic compounds are classified as small (3 and 4 members), normal (5, 6, and 7), medium (8, 9, 10, 11), and large (more than 12) rings.

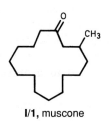

I/1, muscone

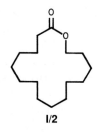

I/2

I/3

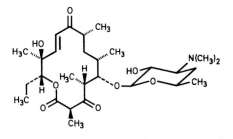

I/4, picromycin

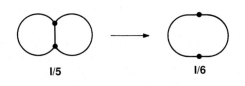

I/5   **I/6**

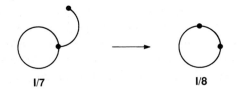

I/7   **I/8**

I/9   **I/10**

Scheme I/1. The principal methods of ring enlargement.

different approaches contains more than hundred methods. Many of them are limited just to one specific type of reaction: The Baeyer-Villiger rearrangement, for instance, allows only the transformation of a cycloalkanone to a lactone containing one additional ring member, an oxygen atom. On the other hand, many methods were developed which can be used in a more general way, to synthesize different types of compounds.

Actually the large number of reaction possibilities can be reduced to only three, which are shown in Scheme I/1. The first one involves the cleavage of the shortest bridge in the bicycle **I/5**. This shortest bridge, representing a single or double bond between the bridgeheads, would be a "zero" bridge, according to IUPAC nomenclature. The bridge can also contain one or more atoms. Depending on the size of the rings of the bicycle and the functional groups placed at, or around, the bridgeheads, the enlargement products, **I/6**, will be different.

The second general way to enlarge a ring is shown by structures **I/7** and **I/8**; the ring is substituted by a single, double or multi–atom side chain, which is placed at a ring atom carrying a suitable functional group. During the ring enlargement process, the side chain is incorporated into the ring. Various types of reaction mechanisms involved in this rearrangement have been discovered.

The final general reaction sequence is the conversion of **I/9** to **I/10**. Two side chains are placed in the same ring at an appropriate distance to each other. With the formation of the new bond, the old one is cleaved. From a mechanistic point of view, pericyclic reactions (electrocyclic and sigmatropic) are of this type.

Although the starting materials, **I/5**, **I/7**, and **I/9**, are different from each other bicyclic intermediates are present in all three. To get a ring enlargement in compounds of type **I/5**, the bridge bond has only to be cleaved. In those of type **I/7**, the functionalized terminal atom of the side chain has to be connected with the ring first. This proposed intermediate is bicyclic and – using our symbols – not different from **I/5**. The true expansion reaction is observed in the next reaction step. Finally, in the third reaction, the transition state between **I/9** and **I/10** is bicyclic and must be cleaved. Thus, if we take the intermediates and transition states into consideration, the number of principal ring enlargement concepts can be reduced to one only, the bicyclic approach, **I/5** → **I/6**.

Although there are many different ways to classify ring enlargement reactions, we have chosen a non-uniform approach as shown in the Table of Contents, because this system allows a better incorporation of the references.

One atom incorporation reactions are discussed in Chapter II; subdivided into carbon, nitrogen, and oxygen incorporation. A few of these reactions are discussed in other sections. Because of their special reactivity most of the three-membered ring compounds used for expansion are combined in Chapter III. Reactions with four-membered intermediates are collected in Chapter IV. Reactions of the type **I/9** → **I/10** will be found in Chapter V and those of **I/7** → **I/8** (see Scheme I/1) in Chapter VII. Bicyclic starting materials will be discussed in the Chapters VIII (cleavage of the zero bridge) and IX (cleavage of an one-atom bridge). The literature on transamidation reactions, including those of β-lactams, is so vast that it takes a special chapter (VI). Thus, the β-lactams are not incorporated into Chapter IV.

Ring enlargement reactions mediated by metals, silicon, or phosphorous are not treated in this survey because of the tremendous amount of material. Rearrangements of bicyclic compounds with a simultaneous contraction and enlargement of the two rings are also excluded.

When we began writing this review, our purpose was to survey ring enlargement methods as complete as possible. However, we found that we had to confine our desire for completeness because of the enormous number of references. The only way to give a clear, concise, and convincing description seemed to be the reaction principles in general and to illustrate them with a selection of striking examples.

References

[1] L. Ruzicka, Helv.Chim.Acta **9**, 1008 (1926).
[2] M. Kerschbaum, Ber.dtsch.chem.Ges. **60B**, 902 (1927).
[3] V. Prelog, J.Chem.Soc. **1950**, 420.
[4] T. G. Back, Tetrahedron **33**, 3041 (1977).
[5] A. C. Cope, M. M. Martin, M. A. McKervey, Quart.Rev. **20**, 119 (1966).
[6] H. Brockmann, W. Henkel, Chem.Ber. **84**, 284 (1951).
[7] H. Brockmann, W. Henkel, Naturwissenschaften **37**, 138 (1950).
[8] S. Masamune, G. S. Bates, J.W. Corcoran, Angew.Chem. **89**, 602 (1977), Angew. Chem.Int.Ed.Engl. **16**, 585 (1977).

II. One-Atom Insertion Procedures

The enlargement of a cyclic organic molecule by one atom is a common reaction, applied almost daily by chemists all over the world. Mostly this atom is carbon, but expansions involving nitrogen and oxygen are also well known. These processes are used industrially on a large scale, especially for the enlargement of carbocycles by one nitrogen atom. The documentation of these reactions in the literature is huge. Thus we cannot review the complete literature, but will only summarize methods. For that reason, we have subdivided this chapter according to the nature of the atoms which are incorporated.

II.1. The One-Carbon Atom Ring Insertion

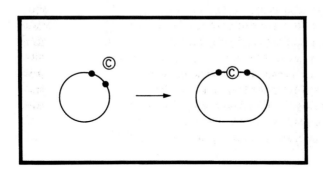

In 1968, an excellent review on "Carbocyclic Ring Expansion Reactions" was published [1]. Most of the reaction discussed there are one carbon atom insertions. Our review will be limited to discussions of newer methods. Well known reactions are summarized only by giving the principal reaction and leading additional references[1]. The principal reactions for one carbon insertion are summarized in Scheme II/1.

---

1) For a review on one carbon ring expansions of bridged bicyclic ketones, see ref. [2].

Pinacol and related rearrangements (Tiffeneau-Demjanow rearrangement, see Scheme II/5)

Wagner-Meerwein rearrangements

Side chain incorporation (see Chapter VII)

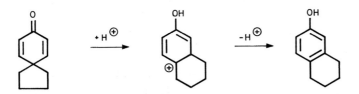

Dienone Phenol rearrangements

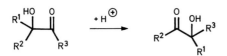

α-Ketol rearrangements

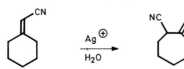

Wittig-Prevóst sequence

Scheme II/1. Types of one carbon insertion reactions.

Pinacol and Related Rearrangements

A large number of one carbon ring expansion procedures are known, depending on the reagents, the reaction conditions, the ring size and its substitution. But, fortunately, the number of fundamental reaction principles is limited. One of these is the pinacol rearrangement. If 1,2-alkanediols are treated with acid, they rearrange to form ketones or aldehydes (**II/1** → **II/5**). The mechanism involves a 1,2-shift of an alkyl substituent (or of hydrogen). More than one rearrangement product can be expected if the substituents at the 1,2-diol, **II/1**, are not identical, Scheme II/2.

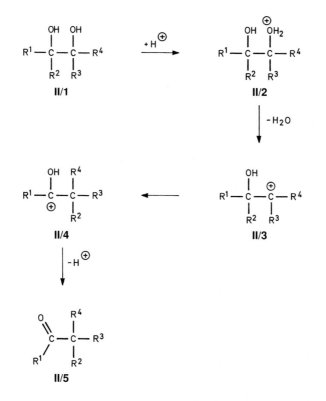

Scheme II/2. The 1,2-shift in a pinacol rearrangement.

A pinacol rearrangement driven by the release of the ring strain in a four-membered ring is shown in Scheme II/3. The exclusive acyl migration from **II/7** to **II/8** is remarkable [3]. Similar reactions have been reported in literature [4].

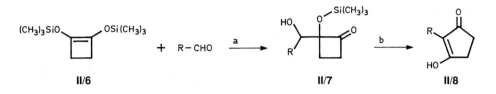

II/6   **II/7**   **II/8**

Scheme II/3. [3]. R = C$_6$H$_5$: a) TiCl$_4$, −78°, 78 %   b) trifluoroacetic acid, 20°, 97 %.

An analogous rearrangement can be observed if one hydroxyl group in compound **II/1** is replaced by another functional group which can place a positive charge at a carbon atom in the neighborhood of C-OH. This type of reaction is called a semipinacol rearrangement, if β-amino alcohols rearrange on treatment with nitrous acid to ketones. A number of one-carbon atom ring expansion reactions follow this pattern.

Wagner-Meerwein Rearrangements

The so-called Wagner-Meerwein[2] rearrangement will be observed if alcohols, especially those substituted by two or three alkyl or aryl groups on the β-carbon atom, are treated with acid. After protonation and loss of water, a 1,2-shift of one of the substituents is observed. Afterwards, the resulting carbocation is stabilized usually by the loss of a hydrogen from the neighboring carbon atom. In a number of cases, substitution products are observed as well as elimination products. A special case of a Wagner-Meerwein reaction is the acid catalyzed conversion of polyspirane **II/9** (Scheme II/4) to the hexacycle, **II/10**, by five ring enlargements one after the other [6].

II/9   **II/10**

Scheme II/4. An example of 1,2-shifts (Wagner-Meerwein rearrangement) [6].

a) TsOH, acetone, H$_2$O, reflux.

---

2) For a review of the Wagner-Meerwein reaction in a fundamental study on equilibria of different ring sizes, see ref. [5].

A small selection of references dealing with ring expansions which follow the Wagner-Meerwein rearrangement is given below:

- From three-membered rings: In protic media, 1-acyl-2-cyclopropene derivatives undergo a ring expansion reaction to cyclobutenols [7]. – Ring expansion of cyclopropylmethanols to fluorinated cyclobutans [8].
- From four-membered rings: An acid-catalyzed transformation has been observed in the conversion of 1-[1-methylsulfinyl-1-(methylthio)alkyl]cyclobutanol to 3-methyl-2-(methylthio)cyclopentanone [9]. – Rearrangement of a β-lactone to a γ-lactone derivative in the presence of magnesiumdibromide [10]. – A borontrifluoride catalyzed cyclobutene to cyclopentene rearrangement [11]. – Ring expansion of a [2+2] photoadduct to a five-membered ring [12].
- From five-membered rings: Synthesis of pyrene derivatives from five-membered ring precursors by ring enlargement [13].
- From six-membered rings: Rearrangement as part of the pseudo-guaianolide to confertin synthesis [14].
- From ten-membered rings: Borontrifluoride catalyzed conversion of germacrane (ten-membered) to humulane (eleven-membered) in 75 % yield [15].

Tiffeneau-Demjanow Rearrangements

The Tiffeneau-Demjanow[3] ring expansion is analogous to the semipinacol rearrangement. It is a homologisation of cyclic ketones. General methods for preparation of the starting 1,2-aminoalcohols from ketones are given in Scheme II/5. They include cyanohydrin, nitromethane, and β-bromoacetic ester approaches. The rearrangement takes place under stereoelectronic control: that bond which is antiperiplanar to the leaving group moves [22].
The reactions of cycloalkanones with diazomethane[1], diazoalkanes, 2-diazocarboxylic acids[4], and trimethylsilyl-diazomethane are also similar to the Tiffeneau-Demjanow rearrangement. These variations are shown in Scheme II/6. Homologation of ketones by diazoalkanes, diazoacetic esters or by the Tiffeneau-Demjanow reaction proceed in good yields although the formation of spiroderivates instead of homologs can be observed. With unsymmetrical ketones, these reactions usually give both types of regioisomers. In order to prevent this uncertainity, better results can be obtained by rearrangement of α-chloroketones. Dechlorination of the final products can be carried out with zinc. An alternative reaction is shown in Scheme II/7. It was used for the trans-

---

3) For a review of the Demjanow and Tiffeneau-Demjanow ring expansions, see ref. [2] [16]. Other references: Comparison of diazomethane and Tiffeneau-Demjanow homologation in the steroid field [17] [18], 9-(aminomethyl)noradamantane [19], 2-adamantanone derivatives [20], in bicyclo[3.3.1]nonan-2-one [21].
4) For reviews see ref. [1] [23] [24].

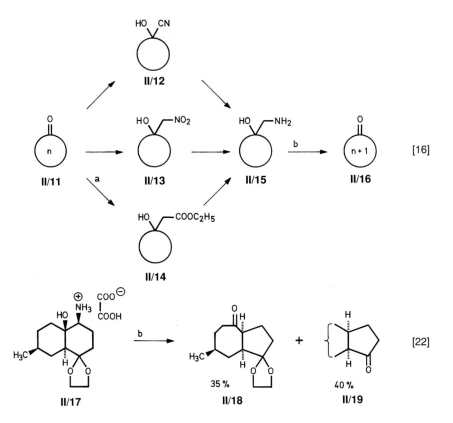

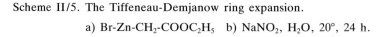

Scheme II/5. The Tiffeneau-Demjanow ring expansion.

a) Br-Zn-CH$_2$-COOC$_2$H$_5$ b) NaNO$_2$, H$_2$O, 20°, 24 h.

formation of cyclododecanone *via* the dibromide **II/35** to cyclotridecanone (**II/39**) [33]. To prevent side reactions especially the formation of oxirane derivatives, the authors suggested that this reaction be performed at −100°, with vigorous stirring, and slow addition of butyllithium [33]. Preparation of dihalo-alcohols, such as **II/35**, can be achieved by reaction of the corresponding ketones with dichloromethyllithium or dibromomethyllithium, followed by hydrolysis. It should be noted that compounds of type **II/35**, prepared from unsymmetrical substituted ketones, can, *a priori* undergo rearrangement in two directions, but rearrangement of the more substituted side is preferred [37]. Further examples are reported in refs. [34] [37] [38] [39] [40].

Another method involves the 1-bromo-2-alkanol derivative, **II/44**, which was prepared from cycloalkanone **II/42** as indicated in Scheme II/7. Compound **II/44** forms a magnesium salt which decomposes to give the 2-phenylcyclo-alkanone **II/46**, enlarged by one carbon atom [35] [41]. The yields are good:

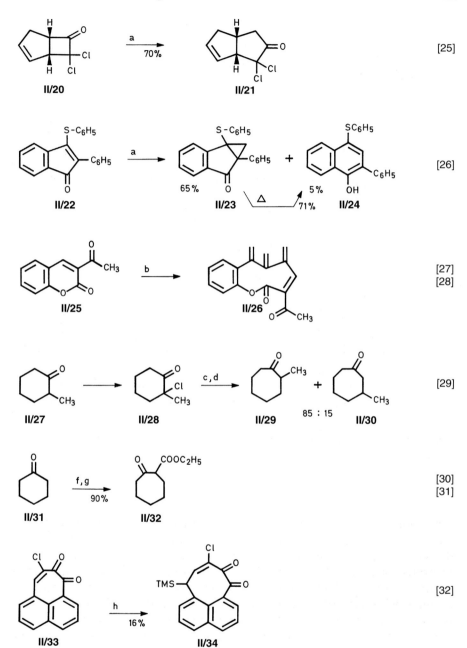

[25]

[26]

[27]
[28]

[29]

[30]
[31]

[32]

Scheme II/6. Ring expansions of cycloalkanones by diazo reagents.

a) CH_2N_2, $(C_2H_5)_2O$ b) $CH_3CH_2N_2$, $(C_2H_5)_2O$ c) $N_2CH_2COOC_2H_5$, $BF_3 \cdot (C_2H_5)_2O$ d) 1. Zn, HOAc 2. Δ, H_2O f) $N_2CH_2COOC_2H_5$, $(C_2H_5)_3OBF_4$, CH_2Cl_2, 0°   g) $NaHCO_3$, H_2O h) $(CH_3)_3SiCH_2N_2$, $BF_3 \cdot (C_2H_5)_2O$, CH_2Cl_2, hexane, −20°.

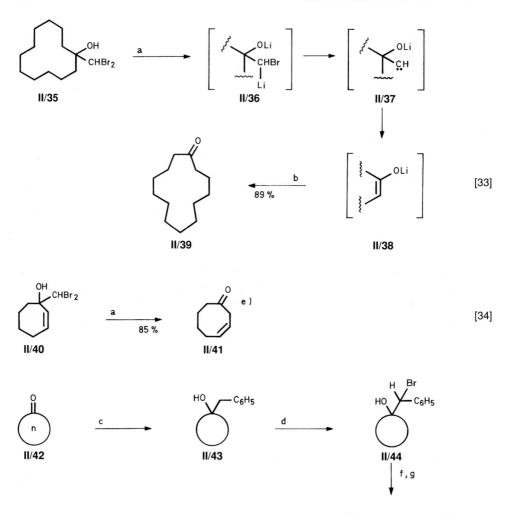

[33]

[34]

[35]

[36]

II/47

R^1 = Alkyl , C$_6$H$_5$

Scheme II/7. Alternative one-carbon ring enlargements.

a) 2 BuLi, −78° b) HCl, H₂O c) C₆H₅CH₂MgCl
d) N-bromosuccinimide, CCl₄ e) The selectivity is better than 98 %
f) *t*-BuMgBr g) benzene, heat h) 3.2 eq. R^2MgBr, THF, −78° → +23°
i) NH₄Cl, H₂O.

II/42 → II/46 *e.g.* n=5: 80 %, n=6: 72 %, n=8: 60 %. – In different reactions ethyl 4-chloromethyl-1,2,3,4-tetrahydro-6-alkyl-2-oxopyrimidine-5-carboxy-lates (**II/47**) are transformed to 4,7-disubstituted ethyl 2,3,6,7-tetrahydro-2-oxo-1*H*-1,3-diazepine-5-carboxylates (**II/49**) using Grignard reagents [36] [36a]. A possible mechanism for this conversion includes the bicycle, **II/48**, Scheme II/7. The alkylation with R^2 takes place after the rearrangement of intermediate **II/48**.

The high reactivity of compounds containing an episulfonium moiety has been used in an one-carbon ring expansion step [42]. This method is explained at the system shown in Scheme II/8. 1-Vinylcyclopentanol is easily prepared from cyclopentanone (**II/50**) and vinyl magnesium bromide. The silylation of the alcohols was carried out with *tert*-butyldimethylsilyloxytriflate (TBDMSOTf). Using trimethylsilylethers instead of TBDMSO-derivatives side reactions are

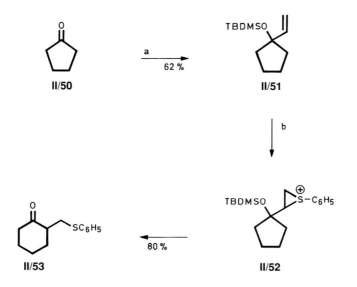

Scheme II/8. An episulfonium ion mediated ring expansion of 1-alkenylcycloalkanols [42].

a) 1. CH₂=CHMgBr, THF 2. TBDMSOTf, 2,6-dimethylpyridine, CH₂Cl₂, 20°
b) C₆H₅SCl, CH₂Cl₂, −78°; AgBF₄, CH₃NO₂, −40°.
TBDMS = *tert*-butyldimethylsilyl.

observed. After treatment of compound **II/51** with C_6H_5SCl the intermediate episulfonium ion **II/52** is destroyed by silver tetrafluoroborate reaction to the six-membered **II/53**.

A further one-carbon atom insertion method is based on the rearrangement of the adducts of cyclic ketones with bis(phenylthio)methyllithium [43]. The

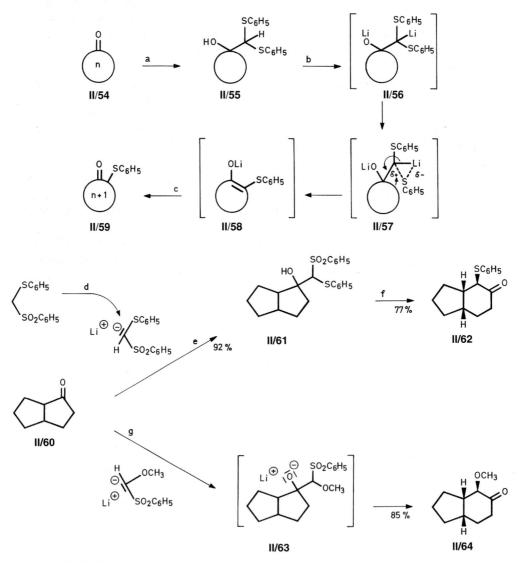

Scheme II/9. Further one-carbon atom insertion methods.

a) LiCH(SC$_6$H$_5$)$_2$, THF, $-78°$ b) 2 CH$_3$Li, $-78°$
c) H$_2$O d) BuLi, THF, $-78°$ e) AlCl(C$_2$H$_5$)$_2$, hexane
f) AlCl(C$_2$H$_5$)$_2$, CH$_2$Cl$_2$, $-78°$
g) 1. 1,2-dimethoxyethane, $-78°$ 2. AlCl(C$_2$H$_5$)$_2$.

reaction principle is shown in Scheme II/9. The products of the expansion are α-phenylthiocycloalkanones **II/59.** A comparison of the results of a number of products formed by this method indicates that a vinyl group migrates faster than an alkyl group and that the more highly substituted alkylgroup migrates preferentially. The yields for the migration step (**II/55 → II/59**) are n=4: 70 %, n=5: 95 %, n=6: 55 %, n=7: 54 % [43]. A copper(I) catalyzed procedure analogous to the transformation **II/54 → II/59** was already published earlier [44]. A treatment of cyclic ketones with tris(methylthio)-methyllithium followed by $CuClO_4 \cdot 4\ CH_3CN$ produces the corresponding ring expanded 2,2-bis(methylthio)cycloalkanones [45].

At the same time the conversion of **II/54 → II/59** (Scheme II/9) was published, an alternative way was found, which is summarized in Scheme II/9. The lithium derivative of (phenylthio)methyl phenyl sulfone adds nearly quantitative into ketones, in the presence of diethylaluminium chloride. The rearrangement (*e.g.* **II/61 → II/62**) proceeds smoothly on treatment of the tertiary alcohol, **II/61**, with an approximately sixfold excess of diethylaluminium chloride [46]. An alternate reagent, the lithium salt of methoxy methyl phenyl sulfone, in a similar reaction yielded, enlarged α-methoxy cycloalkanones. The latter reaction sequence is restricted to the expansion of four- and five-membered rings [46].

A decomposition of β-hydroxyselenids in the presence of thallium dichlorocarbene complex has been used for ring enlargement too, as shown in Scheme II/10, conversion **II/65 → II/67** [47] [48]. – It is reported that a regiospecific

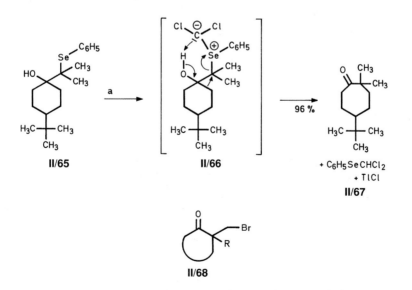

Scheme II/10. A seleno-mediated one-carbon ring expansion [47] [48].

a) $TlOC_2H_5 + CHCl_3 (\rightarrow CCl_2 \cdot TlCl + C_2H_5OH)$, 20°, 8 h.

alkylative ring expansion of 2,2-disubstituted cyclobutanones *via* α-lithioseleno-
xides is possible [49], compare with ref. [50].

The application of α,α-disubstituted cycloalkanones of type **II/68**, Scheme
II/10, for ring enlargement is described in Chapter VII.

Dienone Phenol Rearrangements

The dienone phenol rearrangement[5)] (**II/69** → **II/70**) is another example of a
one-carbon insertion reaction, with the formation of an aromatic system as a
driving force. The reaction is acid catalyzed.

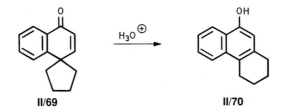

II/69   II/70

α-Ketol Rearrangements

The α-ketol rearrangement[6)] is an isomerization reaction of α-hydroxy ketones
(as well as aldehydes) which takes place under acid as well as base catalysis.
Compound **II/71**, a 17 α-hydroxy-20-ketosteroid, yields, under acid catalysis,
the six-membered isomer **II/72**, and under base catalysis, the mixture of the
isomeric compounds **II/73**, as reviewed in [1].
This reaction has been investigated extensively in D-ring isomerization of
steroids [54] [55]. Only a few examples are known in other systems.

Wittig-Prevóst Method

Aromatic ketones of the α-tetralone type **II/74** can be converted by a Wittig
reaction to compounds of type **II/75**, Scheme II/11. Under Prevóst reaction conditions (AgNO$_3$,I$_2$,CH$_3$OH) two ring enlargement products are formed, **II/77**
and **II/78** (both together in 67 % yield), which by hydrolysis are converted to
the α-cyano ketone **II/79** [56]. This procedure has been applied successfully to

---

5) For reviews of the dienone phenol rearrangement, see refs. [51] [52]. Further reference
[53].
6) For reviews on α-ketol rearrangment, see ref. [1].

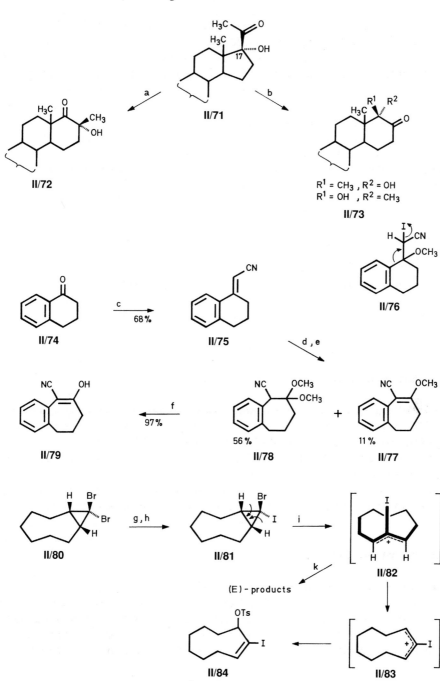

Scheme II/11. a) $H_3O^\oplus$ b) KOH, H_2O
c) NC-CH$_2$-P(O)(OC$_2$H$_5$)$_2$, NaH, 1,2-dimethoxyethane
d) AgNO$_3$, I$_2$, CH$_3$OH e) Icewater f) CF$_3$COOH, CCl$_4$
g) BuLi, THF, $-95°$ h) I$_2$, $-95°$ i) AgOTs, CH$_3$CN k) H$_2$O.

benzannelated seven-membered ring compounds and heterocyclic systems [57]. Presumably the reaction proceeds *via* an intermediate such as **II/76** [57], which has been isolated, and partly transformed into the enlarged product [58].

As demonstrated above, most of the one-carbon insertion reactions are somehow connected with the reactivity of the carbonyl group. This is not true for all cases. Dibromocarbene, prepared by reaction of tetrabromomethane with methyllithium, adds to the double bond of a cycloalkene to give a bicyclic product [59], which, under the influence of a silver salt, forms an enlarged ring [60]. One example is given in Scheme II/11. 9,9-Dibromo-bicyclo[6.1.0]nonane **(II/80)**[7] [61] is converted to its 9-exo-bromo-9-endo-iodo derivative, **II/81**. Ring expansion of **II/81** with silver tosylate in acetonitrile affords exclusively (*Z*)-2-iodo-3-tosyloxy-1-cyclononene **(II/84)** [62] [63] [64]. In the presence of silver perchlorate in 10 % aqueous acetone, a mixture of diastereoisomeric (*Z*), (*E*)-iodo-alcohols was obtained (Scheme II/11). If methyllithium instead of the silver salt is used, the corresponding monocyclic allene will result [59], this can be transformed to a (*Z*)-olefin in sodium liquid ammonia [67]. In a similar reaction sequence, a monohalo bicycle has been converted successfully to the enlarged product [68]. Cyclopropylethers in a bicyclic system can be ring enlarged, *via* a photoinduced single electron transfer promoted opening, in moderate yields [69].

Quite recently a modification of the carbene addition reaction has been published and applied to the synthesis of phoracantholide I **(II/88)** [70]. The silyl enol ether **II/85** prepared from (±)-8-nonanolide underwent addition of chlorocarbene to give the intermediate bicyclic adduct **II/86**, which rearranged into an *E/Z*-mixture of α,β-unsaturated lactones **II/87** by heating. Phoracantholide I **(II/88)** was formed by hydrogenation of the latter, Scheme II/12.

A Vilsmeier-Haack reaction can be used to convert the five-membered isoxazolin-5-one **(II/93)** to the six-membered 4,5-disubstituted 2-dimethylamino-6*H*-1,3-oxazin-6-ones **(II/95)**. The yields are 68–85 % [73]. It has been suggested that the reaction proceeds by attack of the nitrogen atom in **II/93** on the Vilsmeier reagent, followed by ring-opening and cyclisation with hydrogen chloride elimination. Amidines of type **II/95** can be hydrolyzed to give the corresponding 1,3-oxazine-2,6-diones **(II/96)** in 60–85 % yield. A comparable reaction is the transformation of pyrazolo[4.3-*d*]-pyrimidines to pyrimido[5.4-*d*]pyrimidines see ref. [74]. – 2-Cyclobutenylmethanols undergo a 1,2-vinyl shift to 4-chlorocyclopentenes compare ref. [75].

Some miscellaneous ring enlargement reactions are presented in Scheme II/13; two of them are syntheses of cyclobutanones, **II/90** and **II/92** [71] [72].

---

7) A number of other ring enlargement reactions proceed *via* 1,1-dihalocyclopropane intermediates [65] [66].

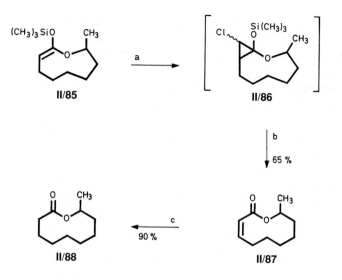

Scheme II/12. Enlargement of a lactone by one-carbon atom [70].

a) CH_2Cl_2, $NaN(Si(CH_3)_3)_2$, pentane; $-25° \rightarrow -20°$, 2 h, then $-20° \rightarrow 0°$
b) toluene, 110°, 15 min   c) H_2/Pd-C, EtOAc.

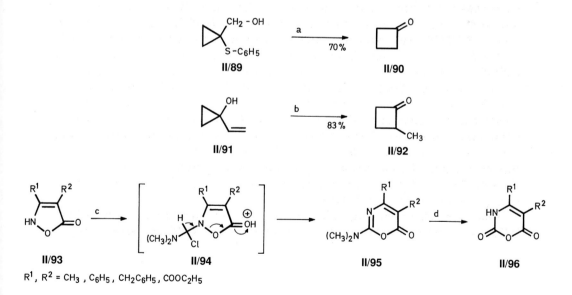

R^1, R^2 = CH_3, C_6H_5, $CH_2C_6H_5$, $COOC_2H_5$

Scheme II/13. Miscellaneous types of one-carbon insertion methods.

a) $HgCl_2$, H_2O, TsOH, $70° \rightarrow 120°$ b) HBr or other electrophiles
c) dimethylformamide, $POCl_3$, CCl_4, heat   d) H_2O.

II.2. Nitrogen Insertion Reactions of Ring Compounds

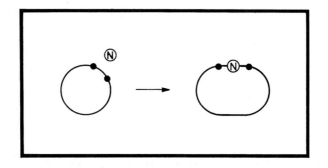

The most important nitrogen insertion reactions are still the Schmidt and the Beckmann rearrangements and their modifications. Yet, a number of other nitrogen insertion reactions are known, and examples will be therefore given and discussed.

The Schmidt Reaction

The Schmidt reaction or Schmidt rearrangement is an insertion method consisting of a reaction between a ketone and hydrazoic acid, in which a cyclic ketone is converted into the corresponding lactam[8]. In Scheme II/14 a mechanistic outline for the Schmidt reaction of 2-methylcyclohexanone (**II/97**) [76] is shown [62] [77] [78]. After addition of hydrazoic acid to the ketone **II/97**, in order to form the protonated azidohydrin **II/98**, a loss of water occurs to give the iminodiazonium ion, **II/99**. By elimination of nitrogen the rearranged iminosalt, **II/100**, is formed, which, after water is added, generates the lactam, **II/101**. It has been shown that in certain cases, the intermediate **II/98** rearranges directly [79]. It is the aryl group which generally migrates with alkyl aryl ketones, except in cases of bulky alkyl groups.

---

8) For reviews of the Schmidt reaction, see ref. [80], with respect to bicyclic ketones, see ref. [81].

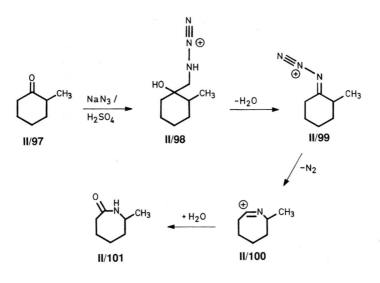

Scheme II/14. Mechanism of the Schmidt reaction with ketones [77].

Schmidt reactions with sodium azide and strong acids, if they occur through tetrahedral reaction intermediates, lead primarily to nitrogen insertion adjacent to methylene rather than methine groups. There are no really satisfactory reasons for preferential methylene migration in this case [81].

The Schmidt reaction with the dienone **II/102** (Scheme II/15) yields the 1,4-thiazepine, **II/103**. Treatment of its dihydroderivate, **II/104**, gives exclusively the enlargement product **II/105**, in which the methylene group migrated [82]. A similar reaction can be observed if the synthetic ergot alkaloid precursors of type **II/106** are treated with *in situ* generated hydrazoic acid. Again no trace of the isomeric lactam can be observed [83].

In the course of the structure elucidation of the natural occurring spermidine alkaloides such as inandeninone **II/108**, (Scheme II/15) the Schmidt reaction played an important role. The "alkaloid" is a nearly 1:1 mixture of two isomers, isolated from *Oncinotis inandensis* Wood et Evans. To make sure that the compounds differed only in the location of the carbonyl group at positions C(12) and C(13), the mixture was treated with sodium azide, sulfuric acid, and chloroform. The product consisted of a mixture of four ring enlarged dilactams (one of them, compound **II/109**, is shown) with nearly equal ratios [84].

Further examples of the Schmidt reaction and of Schmidt type reactions are collected in Scheme II/16.

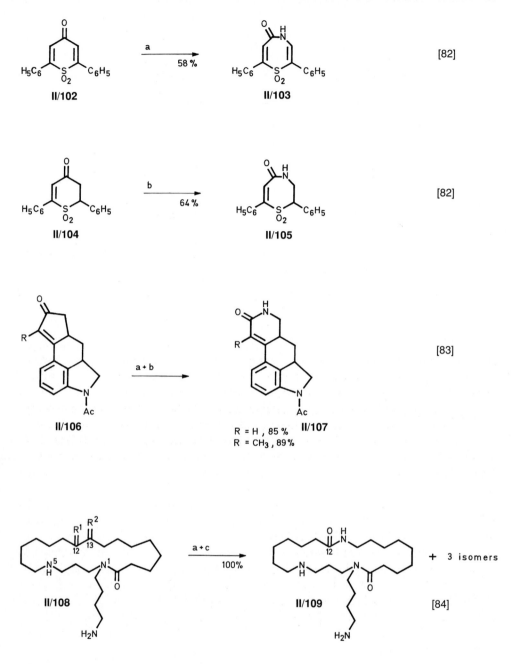

Scheme II/15. Nitrogen insertion by Schmidt reaction.

$R^1 = O, R^2 = H_2$
$R^1 = H_2, R^1 = O$
a) NaN_3, H_2SO_4 b) HOAc, $-65°$ c) $CHCl_3$.

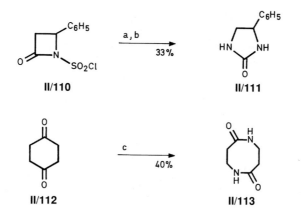

Scheme II/16. Further examples of the Schmidt reaction.

a) NaN$_3$, H$_2$O b) (C$_2$H$_5$)$_2$O, H$_2$O c) HN$_3$, H$_2$SO$_4$, CHCl$_3$.

A variation of the Schmidt type reaction is the rearrangement of an azidocyc-loalkane, which is formed from the addition of hydrazoic acid to an cycloalkene. This reaction was used in the synthesis of muscopyridine (**II/114**), a base isolated from the perfume gland of the musk deer [85]. In this context the reaction of **II/115** as a model compound under the conditions of the Schmidt reaction gave a mixture of two compounds which after dehydrogenation yielded **II/116** and **II/117**. The mechanism can be explained in terms of the migration of different bonds in the precursor.

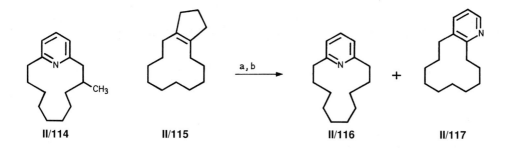

Scheme II/17. Synthesis of an analogue of muscopyridine [85].

a) HN$_3$ b) Pd.

The Beckmann Rearrangement

The Beckmann rearrangement[9] is the second nitrogen insertion reaction which is applied frequently for ring expansions. It takes place when oximes are treated with concentrated sulfuric acid, or PCl_5, or other reagents[10]. In most of the cases, the group which migrates, is the one situated in *anti* position to the hydroxyl group of the oxime. *Syn* group migrations are known, too, and even some, which are not stereospecific [86]. In the latter case it can be assumed that izomerization of the oxime takes place before the migration, and this allows an *anti* migration. In bicyclic systems, the preferential bridgehead migration takes place to the nitrogen atom, but methylene migration has been observed occasionally *e.g.* upon sulfuric acid catalysis [81]. In the first step of the mechanism (Scheme II/18) the hydroxyl group is converted by one of the reagents to a better leaving group. A concerted reaction takes place: Loss of water and migration of the alkyl residue *anti* to the leaving group. The lactam is formed by addition of water to the intermediate carbocation. To illustrate the Beckmann rearrangement, some examples are given in Scheme II/19. – A side reaction is known, called the abnormal Beckmann rearrangement, which consists in the formation of nitriles, *e.g.* ref. [87].

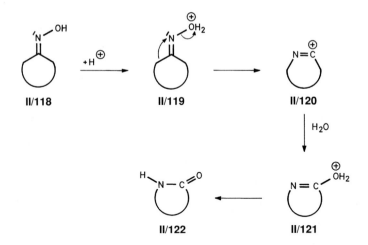

II/118   II/119   II/120

II/122   II/121

Scheme II/18. Mechanism of the Beckmann rearrangement of cycloalkanone.

---

9) For a review with respect to bicyclic oximes see ref. [81].
10) Reagents used in the Beckmann rearrangement are phosphorous [90], formic acid [88], liquid SO_2, P(C_6H_5)_3)-CCl_4, hexamethylphosphorous acid triamide, 2-chloropyridinium fluorosulfonate, SOCl_2, silica gel, P_2O_5-methanesulfonic acid, HCl-HOAc-Ac_2O, polyphosphoric acid [89] as well as trimethylsilyl polyphosphate [88], hydrochloric acid [88], p-toluenesulfonyl chloride [88], trimethylsilyl trifluoromethane sulfonate [93].

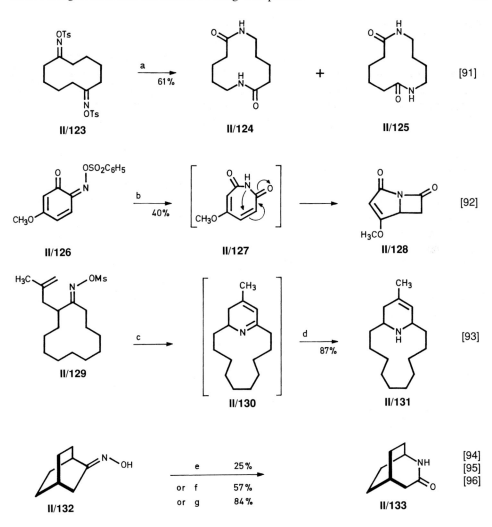

Scheme II/19. The Beckmann rearrangement as a mean for ring expansion.

a) HOAc, NaOAc – H_2O b) pyridine, H_2O, 70°
c) $(CH_3)_3$SiOTf, $CDCl_3$ d) diisobutylaluminiumhydride
e) $C_6H_5SO_2Cl$, NaOH   f) polyphosphoric acid   g) TsCl, pyridine.

The ditosylate of 1,6-cyclodecandion dioxime, **II/123**, after treatment with acetic acid, gave the expected 1:1 mixture of the twelve-membered dilactams, **II/124** and **II/125** [91]. – Instead of the ring enlarged seven-membered ring **II/127**, its ring contraction product, the bicycle **II/128**, can be obtained, if compound **II/126** is heated in aqueous pyridine [92]. – An interesting olefinic cyclization promoted by a Beckmann rearrangement of the oxime mesylate, **II/129**, has been used in a (±)-muscone synthesis [93]. The rearrangement and cyclization product, **II/130**, was reduced to **II/131**, and the latter was transformed to muscone in several steps. – The transformation of the oxime of bicyclo[2.2.2]-octanone (**II/132**) into the lactam, **II/133**, demonstrates that the yield depends very much on the reagents [94] [95] [96].

Several other one-nitrogen-atom expansion reactions are known beside the two nitrogen insertion reactions linked with the names of Schmidt and Beckmann. These reactions are summarized together with references in Schemes II/20 to II/23.

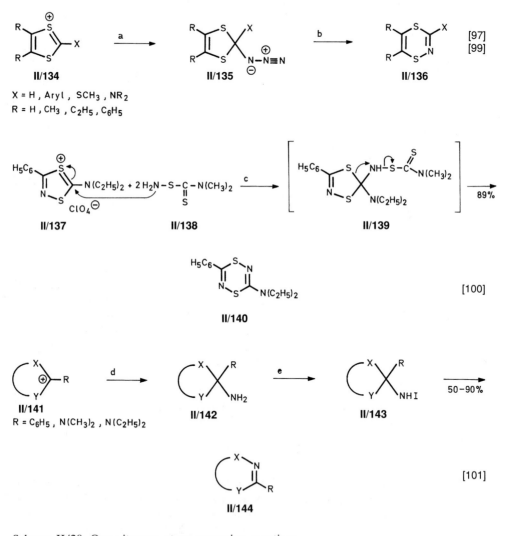

Scheme II/20. One-nitrogen atom expansion reactions.
R = C₆H₅, N(CH₃)₂, N(C₂H₅)₂.
a) NaN₃, CH₃CN   b) heating to m.p., loss of N₂   c) 20°, CH₃CN
d) NH₃, H₂O, CH₃CN   e) I₂.

1,3-Dithiol-2-yl azides (**II/135**), prepared from the corresponding 1,3-dithiolylium salts **II/134** by treatment with sodium azide, are thermally labile. They decompose with an evolution of gas if heated at their melting point. The resulting six-membered dithiazines, **II/136**, are unstable too; they can loose sulfur and give back five-membered compounds, but of different structures [97] [98] [99]. By similar reactions other heterocyclic systems have been synthesized [100] [101] [102], compare [103]. β-Lactams can be prepared from cyclopropanons (**II/145 → II/146 → II/147**) [104] or cyclopropenons (**II/148 → II/149 → II/150 → II/151 + II/152**) [105]; the results are summarized in Scheme II/21.

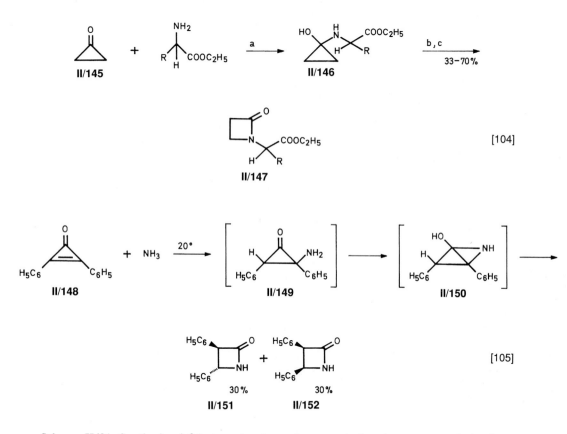

Scheme II/21. Synthesis of β-lactams by ring enlargement of cyclopropanone derivatives.
a) CH$_2$Cl$_2$, −78° b) *t*-BuOCl, CH$_3$CN c) AgNO$_3$.

Under special conditions a number of hydrazine derivatives can be transformed to enlarged ring compounds. A selection of such reactions is collected in Scheme II/22.

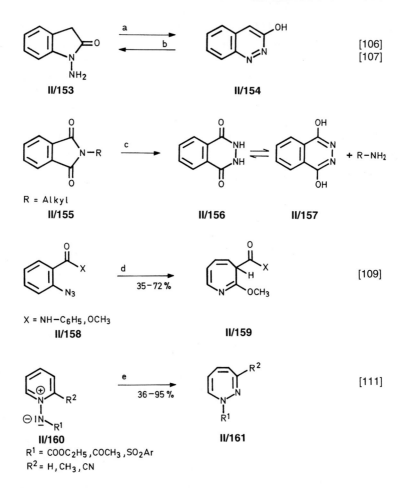

Scheme II/22. One-nitrogen-atom insertion reactions.

a) t-BuOCl, Cl_2, benzene, yield nearly 100 %   b) Zn, H_2SO_4
c) $N_2H_4 \cdot H_2O$, heat   d) hν, CH_3OH e) hν, C_6H_6 or CH_3CN.

1-Aminooxindols (**II/153**) are rearranged to 3-cinnolinols **II/154**, if treated with equimolar amounts of t-butyl hypochlorite [106] [107] [108]. Compound **II/153** is known as "Neber's lactam", and is formed from **II/154** with zinc and sulfuric acid. The mechanism preferred for the **II/153** → **II/154** transformation involves nitrene formation [106]. – As already mentioned, the Gabriel synthesis (**II/155** → **II/157**) is a method for synthesis of primary amines. But the "side" product is the ring enlarged hydrazide, compare Chapter IV.

The thermal and photolytic decomposition of aryl azides in the presence of primary and secondary amines gives 2-amino-3H-azepines [109]. Under similar conditions and in the presence of a variety of different "solvents", azepines

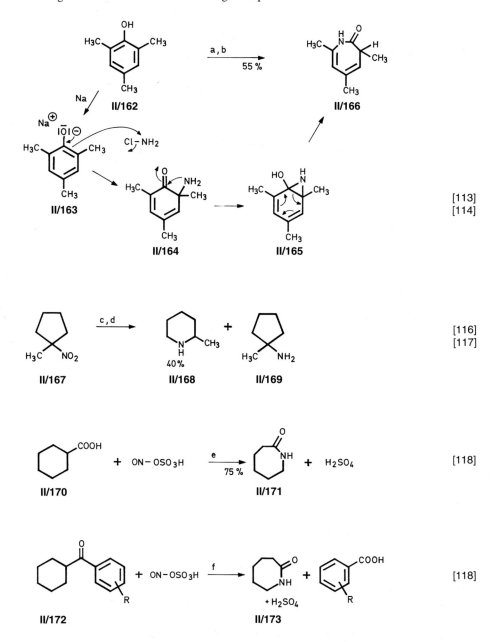

Scheme II/23. Further nitrogen-atom ring expansions.

a) Na b) ClNH$_2$, (C$_2$H$_5$)$_2$O, 125-150° c) LiAlH$_4$, (C$_2$H$_5$)$_2$O, 1 h reflux
d) NaOH, H$_2$O e) 80°, 20 min f) H$_2$O g) t-BuCl, AlCl$_3$, CH$_2$Cl$_2$
h) NaOCl, H$_2$O i) AgCF$_3$COO, CH$_3$OH k) NaOCH$_3$ – NaBH$_4$
m) LiAlH$_4$, (C$_2$H$_5$)$_2$O, reflux n) NaOH, H$_2$O, CH$_3$OH, −15° o) HCl, H$_2$O.

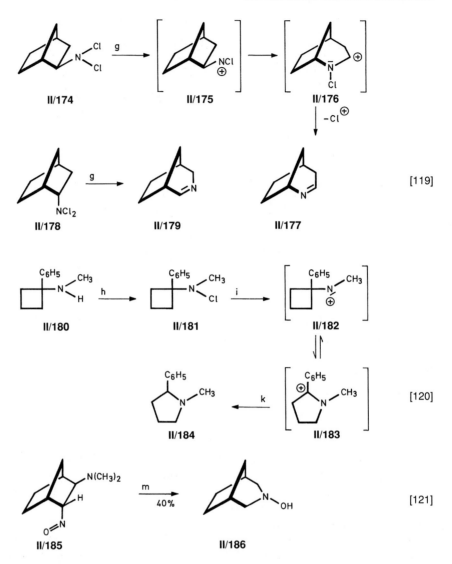

Scheme II/23 (1. continued).

can be synthesized, substituted at position 2 with alkoxy residues (from different alcohols) [109] [110]. Another method of synthesizing 1,2-diazepines, **II/161**, was found when **II/160** was irradiated [111] [112]. – A remarkable one-step ring expansion results if hot solutions of sodio-2,6-dialkylphenoxides in excess 2,6-dialkylphenols, **II/162**, are treated with cold (–70°) etheral chloramine [113] [114]. The mechanistic profile of this reaction presumably involves initial C-amination, followed by thermal rearrangement [115], Scheme II/23. In this Scheme other one-nitrogen-atom incorporation reactions are summarized.

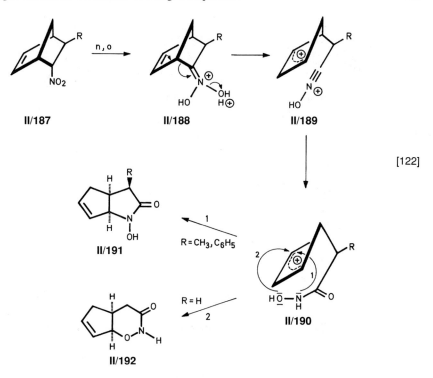

Scheme II/23 (2. continued).

Two reactions of Scheme II/23 should be mentioned. The two epimers of 2-(N,N-dichloroamino)norbornane, **II/174** and **II/178**, when treated with AlCl₃-CH₂Cl₂, gave two different ring enlargement products, **II/177** and **II/179**, respectively [119]. The incorporation of nitrogen into the ring is a stereo-controlled migration reaction.

The other reaction of Scheme II/23, worth mentioning is the transformation of the four-membered **II/180** to the five-membered **II/184** [120]. The generation of a nitrenium ion adjacent to a small ring in **II/182** results in expansion of the small to the larger nitrogen containing ring. The reaction has something in common with the Demjanow rearrangement in carbonium ion chemistry. The equilibrium between the four- and five-membered intermediates has been deduced from the results obtained by reduction of the total reaction mixture: 36 % of **II/184** and 24 % of **II/180**.

Treatment of the 5-nitronorbornenes, **II/187**, (Scheme II/23) with base followed by acid gives, depending on the nature of additional substituents, two kinds of annelated ring enlarged products, **II/191** and **II/192**, respectively. The rearrangement probably proceeds by a double protonated nitronate ion. Striking details of the mechanism are the ring opening and elimination of water, followed by addition of water in the opposite sense to give the protonated nitrile oxide and ring closure to **II/191** and **II/192** [122].

Reactions of phenylnitrenes generated *in situ* were used for the synthesis of acridones [123]; for synthesis of azacycloheptatriene derivatives see ref. [124].

II.3. Oxygen Insertion Reactions

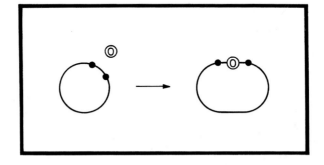

It has been shown that the peracid oxidation of cyclic ketones, **II/193**, is the best synthetic method for lactone, **II/197**, formation by oxygen insertion. This reaction is called the Baeyer-Villiger rearrangement (or Baeyer-Villiger oxidation)[11]. The list of oxidizing agents is fairly long: organic peracids, such as peracetic, perbenzoic, *m*-chloroperbenzoic, pertrifluoric acids; inorganic oxidizing agents, like H_2SO_5, ceric ammonium nitrate, or ceric ammonium sulfate, vinyl acetate/O_3 + heat [131], *m*-chloroperbenzoic acid plus 2,2,6,6-tetramethyl-piperidine hydrochloride [132]; and *m*-chloroperbenzoic acid together with trifluoroacetic acid [126].

Scheme II/24 gives the proposed mechanism [133]. The central intermediate, **II/195**, is tetrahedral. From studies of a series of unsymmetrical substituted ketones, the relative ease of migration of various groups has been found to be: *tert*-alkyl (best) > cyclohexyl, sec-alkyl, benzyl, phenyl > prim-alkyl > methyl [134]. To illustrate this method, some examples of the Baeyer-Villiger rearrangement are given in Scheme II/25[12].

---

11) For reviews (application to bicyclic ketones): [125]. Some more recent examples are shown in [126] [127] [128] [129] [130].

12) In the synthesis of the following lactones, the oxygen atom has been introduced by Baeyer-Villiger rearrangement, *e.g.* (±)-phoracantholide I [137], (±)-lactone antibiotic A26771B [138], exaltone® [139].

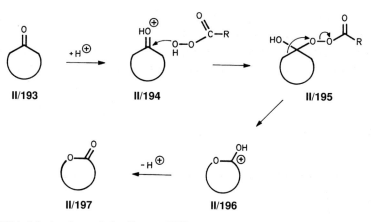

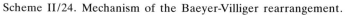

Scheme II/24. Mechanism of the Baeyer-Villiger rearrangement.

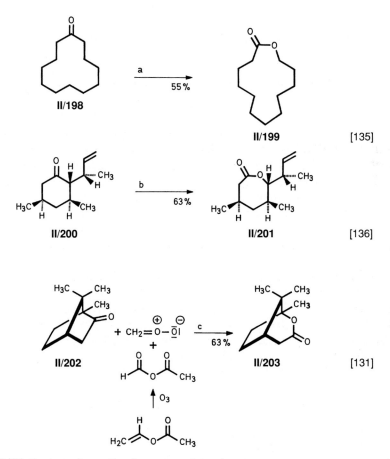

Scheme II/25. Lactone formation by oxygen insertion.

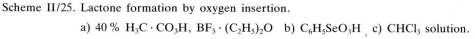

a) 40 % $H_3C \cdot CO_3H$, $BF_3 \cdot (C_2H_5)_2O$ b) $C_6H_5SeO_3H$ c) $CHCl_3$ solution.

An alternative method to the Baeyer-Villiger reaction is that of oxidation with formaldehyde oxide. The latter can be generated by treatment of vinyl acetate with ozone, which gives, beside the wanted products, the mixed anhydride of acetic and formic acid. If a ketone is added to the ozonide mixture, formaldehyde oxide can be trapped to yield the corresponding ozonide, which on thermal decomposition forms the lactone and (or) the corresponding olefinic carboxylic acid. The transformation of camphor (**II/202**) to the lactone **II/203** by this method has been achieved in 63 % yield [131].

References

[1] C. D. Gutsche, D. Redmore "Carbocyclic Ring Expansion Reactions", Academic Press, New York 1968.
[2] G. R. Krow, Tetrahedron **43**, 3 (1987).
[3] E. Nakamura, I. Kuwajima, J.Am.Chem.Soc. **99**, 961 (1977).
[4] I. Kuwajima, I. Azegami, Tetrahedron Lett. **1979**, 2369.
[5] R. P. Kirchen, T. S. Sorensen, K. E. Wagstaff, J.Am.Chem.Soc. **100**, 5134 (1978).
[6] L. Fitjer, D. Wehle, M. Noltemeyer, E. Egert, G. M. Sheldrick, Chem.Ber. **117**, 203 (1984).
[7] M. Vincens, C. Dumont, M. Vidal, Tetrahedron Lett. **27**, 2267 (1986).
[8] S. Kanemoto, M. Shimizu, H. Yoshioka, Bull.Chem.Soc.Jpn. **62**, 2024 (1989).
[9] K. Ogura, M. Yamashita, M. Suzuki, G.-i. Tsuchihashi, Chem.Lett. **1982**, 93.
[10] T. H. Black, J. A. Hall, R. G. Sheu, J.Org.Chem. **53**, 2371 (1988).
[11] C. Kaneko, Y. Momose, T. Maeda, T. Naito, M. Somei, Heterocycles **20**, 2169 (1983).
[12] G. L. Lange, C. P. Decicco, J. Willson, L. A. Strickland, J.Org.Chem. **54**, 1805 (1989).
[13] T. Kimura, M. Minabe, K. Suzuki, J.Org.Chem. **43**, 1247 (1978).
[14] J. A. Marshall, R. H. Ellison, J.Am.Chem.Soc. **98**, 4312 (1976).
[15] V. Enev, E. Tsankova, Tetrahedron Lett. **29**, 1829 (1988).
[16] P.A. S. Smith, D. R. Baer, Org. Reactions **11**, 157 (1960).
[17] J. B. Jones, P. Price, Tetrahedron **29**, 1941 (1973).
[18] G. Haffer, U. Eder, G. Neef, G. Sauer, R. Wiechert, Liebigs Ann.Chem. **1981**, 425.
[19] M. Nakazaki, K. Naemura, M. Hashimoto, J.Org.Chem. **48**, 2289 (1983).
[20] R. K. Murray, T. M. Ford, J.Org.Chem. **44**, 3504 (1979).
[21] T. Momose, O. Muraoka, N. Shimada, C. Tsujimoto, T. Minematsu, Chem.Pharm.Bull. **37**, 1909 (1989).
[22] T. Weller, D. Seebach, R. E. Davies, B. B. Laird, Helv.Chim.Acta **64**, 736 (1981).
[23] C. D. Gutsche, Org. Reactions **8**, 364 (1954).
[24] E. Müller, M. Bauer, Liebigs Ann.Chem. **654**, 92 (1962).
[25] A. E. Greene, J.-P. Deprés, J.Am.Chem.Soc. **101**, 4003 (1979).
[26] K. Buggle, U. N. Ghógáin, D. O'Sullivan, J.Chem.Soc. Perkin Trans. I **1983**, 2075.
[27] F. M. Dean, B. K. Park, J.Chem.Soc., Chem.Commun. **1975**, 142.
[28] R. Clinging, F. M. Dean, L. E. Houghton, J.Chem.Soc. C **1971**, 66.
[29] V. Dave, E.W. Warnhoff, J.Org.Chem. **48**, 2590 (1983).
[30] W. L. Mock, M. E. Hartman, J.Am.Chem.Soc. **92**, 5767 (1970).
[31] W. L. Mock, M. E. Hartman, J.Org.Chem. **42**, 459 (1977).
[32] J. Ikuina, K. Yoshida, H. Tagata, S. Kumakura, J. Tsunetsugu, J.Chem.Soc. Perkin Trans. I **1989**, 1305.
[33] H. Taguchi, H. Yamamoto, H. Nozaki, J.Am.Chem.Soc. **96**, 6510 (1974).
[34] H. Taguchi, H. Yamamoto, H. Nozaki, Bull.Chem.Soc.Jpn. **50**, 1592 (1977).

[35] A. J. Sisti, J.Org.Chem. **33**, 3953 (1968).

[36] D. A. Claremon, S. A. Rosenthal, Synthesis **1986**, 664.

[36a] D. A. Claremon, D. E. McClure, J. P. Springer, J. J. Baldwin, J.Org.Chem. **49**, 3871 (1984).

[37] H. Taguchi, H. Yamamoto, H. Nozaki, Tetrahedron Lett. **1976**, 2617.

[38] J. Villieras, P. Perriot, J. F. Normant, Synthesis **1979**, 968.

[39] K. C. Nicolaou, R. L. Magolda, D. A. Claremon, J.Am.Chem.Soc. **102**, 1404 (1980).

[40] L. A. Paquette, R. Kobayashi, M. A. Kesselmayer, J. C. Gallucci, J.Org.Chem. **54**, 2921 (1989).

[41] A. J. Sisti, J.Org.Chem. **33**, 453 (1968).

[42] S. Kim, J. H. Park, Tetrahedron Lett. **30**, 6181 (1989).

[43] W. D. Abraham, M. Bhupathy, T. Cohen, Tetrahedron Lett. **28**, 2203 (1987).

[44] T. Cohen, D. Kuhn, J. R. Falck, J.Am.Chem.Soc. **97**, 4749 (1975).

[45] S. Knapp, A. F. Trope, R. M. Ornaf, Tetrahedron Lett. **1980**, 4301.

[46] B. M. Trost, G. K. Mikhail, J.Am.Chem.Soc. **109**, 4124 (1987).

[47] J. L. Laboureur, A. Krief, Tetrahedron Lett. **25**, 2713 (1984).

[48] D. Labar, J. L. Laboûreur, A. Krief, Tetrahedron Lett. **23** 983 (1982).

[49] R. C. Gadwood, J.Org.Chem. **48**, 2098 (1983).

[50] S. Halazy, F. Zutterman, A. Krief, Tetrahedron Lett. **23**, 4385 (1982).

[51] B. Miller, Acc.Chem.Res. **8**, 245 (1975).

[52] B. Miller, Mechanism of Molecular Migrations **1**, 247 (1968).

[53] B. Hagenbruch, S. Hünig, Chem.Ber. **116**, 3884 (1983).

[54] E. Keinan, Y. Mazur, J.Org.Chem. **43**, 1020 (1978).

[55] N. L. Wendler, D. Taub, R.W. Walker, Tetrahedron **11**, 163 (1960).

[56] M. Geier, M. Hesse, Synthesis **1990**, 56.

[57] M. S. El-Hossini, K. J. McCullough, R. McKay, G. R. Proctor, Tetrahedron Lett. **1986**, 3783.

[58] M. Geier, H. Zürcher, M. Hesse, 1990 to be published.

[59] K. G. Untch, D. J. Martin, N.T. Castellucci, J.Org.Chem. **30**, 3572 (1965).

[60] P. S. Skell, S. R. Sandler, J.Am.Chem.Soc. **80**, 2024 (1958).

[61] J. P. Marino, L. J. Browne, Tetrahedron Lett. **1976**, 3241.

[62] H. J. J. Loozen, W. M. M. Robben, H. M. Buck, Rec.Trav.Chim. Pays-Bas **95**, 245 (1976).

[63] H. J. J. Loozen, W. M. M. Robben, H. M. Buck, Rec.Trav.Chim. Pays-Bas **95**, 248 (1976).

[64] H. J. J. Loozen, W. M. M. Robben, T. L. Richter, H. M. Buck, J.Org.Chem. **41**, 384 (1976).

[65] E. Vogel, Angew.Chem. **72**, 4 (1960).

[66] P.W. Hickmott, Tetrahedron **40**, 2989 (1984).

[67] S. N. Moorthy, R. Vaidyanathaswamy, D. Devaprabhakara, Synthesis **1975**, 194.

[68] G. H. Whitham, M. Wright, J.Chem.Soc., Chem.Commun. **1967**, 294.

[69] P. G. Gassman, S. J. Burns, J.Org.Chem. **53**, 5576 (1988).

[70] E. Fouque, G. Rousseau, Synthesis **1989**, 661.

[71] B. M. Trost, W. C. Vladuchick, Synthesis **1978**, 821.

[72] H. H. Wasserman, R. E. Cochoy, M. S. Baird, J.Am.Chem.Soc. **91**, 2376 (1969).

[73] E. M. Beccalli, A. Marchesini, H. Molinari, Tetrahedron Lett. **27**, 627 (1986).

[74] S. Senda, K. Hirota, T. Asao, Y. Yamada, Tetrahedron Lett. **26**, 2295 (1978).

[75] J. A. Miller, G. M. Ullah, J.Chem.Soc., Chem.Commun. **1982**, 874.

[76] H. Shechter, J. C. Kirk, J.Am.Chem.Soc. **73**, 3087 (1951).

[77] R. C. Elderfield, E.T. Losin, J.Org.Chem. **26**, 1703 (1961).

[78] R. D. Bach, G. J. Wolber, J.Org.Chem. **47**, 239 (1982).

[79] L. E. Fikes, H. Shechter, J.Org.Chem. **44**, 741 (1979).

[80] G. Koldobskii, G. F. Teveschchenko, E. S. Gerasimova, L. I. Bagal, Russ.Chem.Rev. **40**, 835 (1971).

[81] G. R. Krow, Tetrahedron **37**, 1283 (1981).
[82] W. Ried, H. Bopp, Synthesis **1978**, 211.
[83] D. C. Horwell, D. E. Tupper, W. H. Hunter, J.Chem.Soc. Perkin Trans. I **1983**, 1545.
[84] M. M. Badawi, K. Bernauer, P.v.d.Broek, D. Gröger, A. Guggisberg, S. Johne, I. Kompíš, F. Schneider, H.-J. Veith, M. Hesse, H. Schmid, Pure Appl.Chem. **33**, 81 (1973).
[85] K. Biemann, G. Büchi, B. H. Walker, J.Am.Chem.Soc. **79**, 5558 (1957).
[86] P.T. Lansbury, N. R. Mancuso, Tetrahedron Lett. **1965**, 2445.
[87] B. Amit, A. Hassner, Synthesis **1978**, 932.
[88] I. Ganboa, C. Palomo, Synth.Commun. **13**, 941 (1983).
[89] A. Guy, J.-P. Guetté, G. Lang, Synthesis **1980**, 222.
[90] J. March, "Advanced Organic Chemistry", 3. Ed., John Wiley & Sons, New York 1985.
[90a] R. Graf, Liebigs Ann.Chem. **661**, 111 (1963).
[91] M. Rothe, Chem.Ber. **95**, 783 (1962).
[92] N. Hatanaka, H. Ohta, O. Simamura, M. Yoshida, J.Chem.Soc., Chem.Commun. **1971**, 1364.
[93] S. Sakane, K. Maruoka, H. Yamamoto, Tetrahedron Lett. **24**, 943 (1983).
[94] H. K. Hall, J.Am.Chem.Soc. **82**, 1209 (1960).
[95] G. Reinisch, H. Bara, H. Klare, Chem.Ber. **99**, 856 (1966).
[96] K.-i. Morita, Z. Suzuki, J.Org.Chem. **31**, 233 (1966).
[97] E. Fanghänel, K.-H. Kühnemund, A. M. Richter, Synthesis **1983**, 50.
[98] J. Nakayama, M. Ochiai, K. Kawada, M. Hoshino, J.Chem.Soc. Perkin Trans. I **1981**, 618.
[99] J. Nakayama, A. Sakai, A. Tokiyama, M. Hoshino, Tetrahedron Lett. **24**, 3729 (1983).
[100] K. Yonemoto, I. Shibuya, K. Honda, Bull.Chem.Soc.Jpn. **61**, 2232 (1988).
[101] K. Yonemoto, I. Shibuya, Chem.Lett. **1989**, 89.
[102] E. Fanghänel, J.prakt.Chemie **318**, 127 (1976).
[103] J. Nakayama, H. Fukushima, R. Hashimoto, M. Hoshino, J.Chem.Soc., Chem. Commun. **1982**, 612.
[104] H. H. Wasserman, E. Glazer, J.Org.Chem. **40**, 1505 (1975).
[105] F. Toda, T. Mitote, K. Akagi, Bull.Chem.Soc.Jpn. **42**, 1777 (1969).
[106] H. E. Baumgarten, W. F. Wittman, G. J. Lehmann, J.Heterocycl.Chem. **6**, 333 (1969).
[107] H. E. Baumgarten, P. L. Creger, R. L. Zey, J.Am.Chem.Soc. **82**, 3977 (1960).
[108] H. E. Baumgarten, P. L. Creger, J.Am.Chem.Soc. **82**, 4634 (1960).
[109] R. K. Smalley, W. A. Strachan, H. Suschitzky, Synthesis **1974**, 503.
[110] A. C. Mair, M.F. G. Stevens, J.Chem.Soc. C **1971**, 2317.
[111] J. Streith, J.-M. Cassal, Bull.Soc.Chim.France **1969**, 2175.
[112] M. Nastasi, J. Streith, Bull.Soc.Chim.France **1973**, 630.
[113] L. A. Paquette, J.Am.Chem.Soc. **84**, 4987 (1962).
[114] L. A. Paquette, J.Am.Chem.Soc. **85**, 3288 (1963).
[115] L. A. Paquette, W. C. Farley, J.Am.Chem.Soc. **89**, 3595 (1967).
[116] G. E. Lee, E. Lunt, W. R. Wragg, H. J. Barber, Chem. & Industry **1958**, 417.
[117] H. J. Barber, E. Lunt, J.Chem.Soc. **1960**, 1187.
[118] C. Ferri, "Reaktionen der organischen Synthese", Thieme-Verlag, Stuttgart 1978.
[119] P. Kovacíc, M. K. Lowery, P. D. Roskos, Tetrahedron **26**, 529 (1970).
[120] P. G. Gassman, A. Carrasquillo, Tetrahedron Lett. **1971**, 109.
[121] S.-C. Chen, Tetrahedron Lett. **1972**, 7.
[122] W. E. Noland, R. B. Hart, W. A. Joern, R. G. Simon, J.Org.Chem. **34**, 2058 (1969).
[123] D. G. Hawkins, O. Meth-Cohn, J.Chem.Soc. Perkin Trans I **1983**, 2077.
[124] S. Murata, T. Sugawara, H. Iwamura, J.Chem.Soc., Chem.Commun. **1984**, 1198.
[125] G. R. Krow, Tetrahedron **37**, 2697 (1981).
[126] S. S. C. Koch, A. R. Chamberlin, Synth.Commun. **19**, 829 (1989).
[127] M. J. Taschner, D. J. Black, J.Am.Chem.Soc. **110**, 6892 (1988).
[128] S. Bienz, M. Hesse, Helv.Chim.Acta **70**, 1333 (1987).

[129] M. Matsumoto, H. Kobayashi, Heterocycles **24**, 2443 (1986).

[130] R. Noyori, T. Sato, H. Kobayashi, Bull.Chem.Soc.Jpn. **56**, 2661 (1983).

[131] R. Lapalme, H.-J. Borschberg, P. Saucy, P. Deslongchamps, Can.J.Chem. **57**, 3272 (1979).

[132] J. A. Cella, J. P. McGrath, J. A. Kelley, O. ElSoukkary, L. Hilpert, J.Org.Chem. **42**, 2077 (1977).

[133] R. Criegee, Liebigs Ann.Chem. **560**, 127 (1948).

[134] H. O. House, "Modern Synthetic Reactions", 2. Ed., Benjamin, Menlo Park CA, 1972.

[135] B. D. Mookherjee, R.W. Trenkle, R. R. Patel, J.Org.Chem. **37**, 3846 (1972).

[136] P. A. Grieco, Y. Yokoyama, S. Gilman, Y. Ohfune, J.Chem.Soc., Chem.Commun. **1977**, 870.

[137] T. Ohnuma, N. Hata, N. Miyachi, T. Wakamatsu, Y. Ban, Tetrahedron Lett. **27**, 219 (1986).

[138] C.T. Walsh, Y.-C. J. Chen, Angew.Chem. **100**, 342 (1988); Angew. Chem. Int. Ed. Engl. **27**, 333 (1988).

[139] C. Fehr, Helv.Chim.Acta **66**, 2512 (1983).

III. The Three-membered Ring – a Building Element for Ring Enlargement Reactions

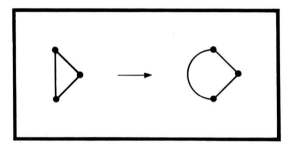

Because of their ring strain, three-membered rings are good starting materials for ring expansion reactions. In this chapter, two types of ring enlargement reactions will be discussed in which three-membered rings are involved. One of the rings is the heterocycle aziridine and its derivatives and the other is cyclopropane [1] [2].

Aziridine Derivatives

Aziridine and its derivatives are reactive ring enlargement reagents. Three atoms, including the nitrogen atom, are involved in the expansion step. 3-Dimethylamino-2H-azirine is a cyclic three-membered amidine, containing electrophilic as well as nucleophilic centers. A number of interesting and useful synthetic reactions have been discovered in which 3-amino-2H-azirines are applied to different substrates [3] [4]. Most of the reactions were carried out with 3-dimethylamino-2,2-dimethyl-2H-azirine (**III/2**). The treatment of **III/2** with the malonimide, 3,3-diethylazetidine-2,4-dione (**III/1**), in the presence of isopropanol leads in high yield to the seven-membered heterocycle, **III/5** [5], Scheme III/1. For this reaction it is necessary that reagent **III/1** be sufficiently acidic to protonate the basic nitrogen atom, N(1), of the azirine. The remaining anion should be a nucleophile which attacks C(3) of the protonated **III/2** to give the enlarged heterocycle, **III/5**, *via* **III/3** and the bicyclic intermediate, **III/4**.

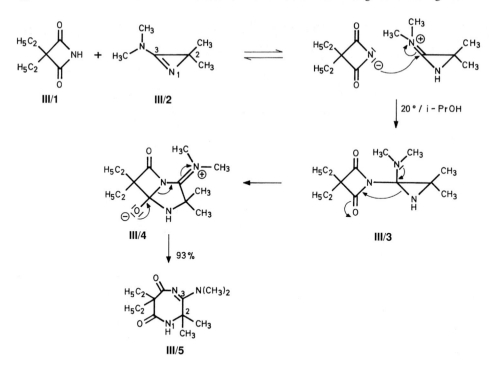

Scheme III/1. The "blowing up" of malonimide by 3-dimethylamino-2*H*-azirine [5].

Depending on the substrates, a large variety of heterocyclic systems with different ring sizes can be built up. In Scheme III/2 some of these are collected. Saccharine (**III/6**), yielded the benzannelated eight-membered ring compound, **III/7** [7] in reacting with **III/2**. Phthalimide (**III/8**), gave the corresponding ring enlarged product, **III/9**, which on recrystallization cyclized by a transannular reaction to **III/10** [8]. Transannular reactions can be expected in the enlargement products of azirines, because they are often polyfunctional medium-sized heterocycles. If the 4,4-dimethyl-3-isoxazolidinone **III/11** is treated with the azirine, **III/2**, [6] [9] it is transformed to **III/12**. The benzothiadiazine derivative, **III/13**, can be converted to the nine-membered **III/14**, again in a nearly quantitative yield [10]. Diphenylcyclopropenone [11] as well as diphenylcyclopropenethione react with **III/2** to form the pyridones, **III/15**, [12] and pyridinethiones, **III/17**, [13], respectively, Scheme III/3. Probably in both reactions the two intermediates, **III/16** and **III/18**, are involved.

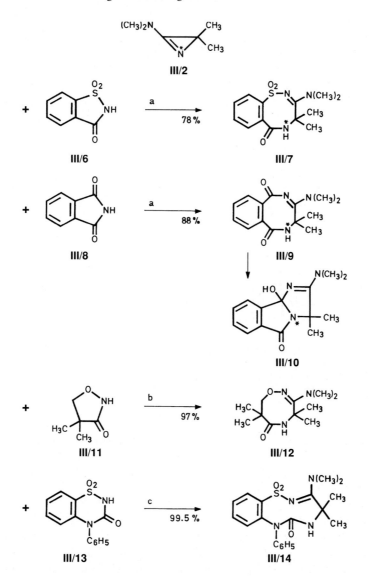

Scheme III/2. Azirines as ring enlargement units. The stared nitrogen atoms were caused from ^{15}N by labelling experiments [6].

a) Dimethylformamide, 0°   b) CH_3CN, 20°, 24 h   c) $CHCl_3$, −15°.

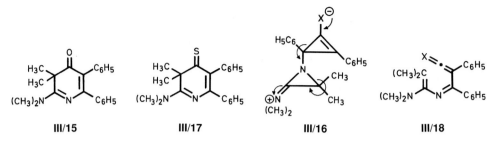

III/15   **III/17**   **III/16**   **III/18**

Scheme III/3. X = S or O

The application of 3-amino-2*H*-azirines as amino acid equivalents and their use in the synthesis of cyclic dipeptides presumably includes ring enlargement steps [14]. The formation of 12- or 15-membered cyclodepsipeptides from open chain precursors possibly includes ring enlargement steps [3] [15].

The treatment of aziridines with nucleophiles having a latent electrophilic site is known to be a method for constructing heterocyclic compounds, Scheme III/4 [16] [17]. Aziridine (**III/19**) itself reacts with α-amino acid esters to give piperazinones (**III/20**) [18] [19], while aziridine with α-mercapto ketones form thiazines [11] [16] [17] and with malonate ester substituted pyrrolidones [20] [21]. In the presence of sodium methylate 3-mercapto-propanoates react with azirines to give thialactams (**III/19** → **III/32**) [22]. A tentative mechanism of this reaction is proposed in Scheme III/5. The reaction of substituted aziridines of type **III/21** (as well as the corresponding NH derivatives) with optically active α-amino acids lead to the diastereoisomeric ring enlarged products, **III/22** and **III/23** (56 % yield of an approx. 1:1 mixture) [23], Scheme III/4. The reaction of the aziridine derivative, **III/24**, with dimethyl acetylenedicarboxylate has been investigated by two research groups. Both isomeric aziridines, **III/24**, the *cis-* and the *trans*-isomers, lead to the same result, a crystalline dihydropyrrole derivative in excellent yield, if the components are heated in a sealed tube at 120°. In the first published paper [24] the structure **III/30** was assigned to the compound. Reinvestigation of this thermolysis reaction led to assignment of the structure **III/29** for the main product [25]. The proposed mechanism [25] is outlined in Scheme III/4. By separate experiments, it was shown that the intermediate **III/26** reacts in a [2+2] cycloaddition to form the bicyclic compound, **III/27**. The latter rearranges thermally *via* ring opening to **III/28** and ring closing to the dihydropyrrole derivative, **III/29**. This transformation is characterized by an interesting set of alternate expansion, cleavage and, finally, expansion reactions.

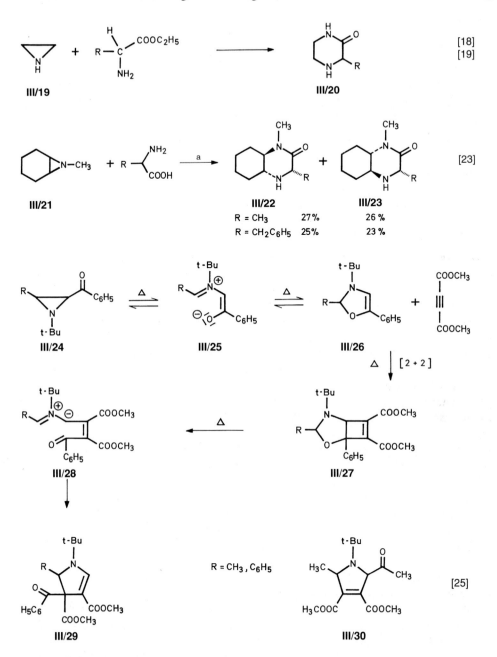

Scheme III/4. Ring enlargement reactions with aziridines.

a) NH$_4$Cl, H$_2$O, reflux 18 h.

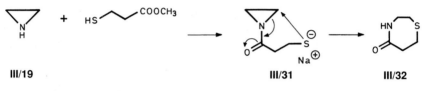

III/19   **III/31**   **III/32**

Scheme III/5. a) CH₃OH, CH₃ONa.

The expected acid chloride was not observed on treatment of the N-*tert*-butyl-aziridine, **III/33**, with a chlorinating reagent such as thionylchloride or oxalylchloride. Instead, the α-chloro-β-lactam, **III/35**, was formed by neighbouring group participation of the nitrogen atom. The mild reaction conditions, the good yields, and the stereospecifity make this ring expansion a fine method for potentially useful β-lactams [26], Scheme III/6.

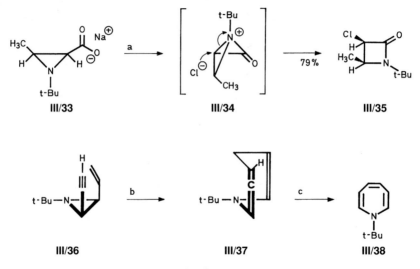

III/33   **III/34**   **III/35**

III/36   **III/37**   **III/38**

Scheme III/6. a) (COCl)₂ or SOCl₂   b) [3.3] sigmatropic shift, sealed tube 80 – 130°
c) [1.3] shift.

A final example of the ring enlargement with aziridine derivatives is the rearrangement of *cis*-N-*tert*-butyl-2-ethynyl-3-vinylaziridine (**III/36**). By a thermal reaction, **III/36**, was converted to N-*tert*-butyl-1*H*-azepine, **III/38**, including a [3.3] sigmatropic rearrangement followed by a [1.3] hydrogen shift [27]. The transformation, **III/36** to **III/38**, has been classified as an example of an expansion mediated by a three-membered ring. Because of the mechanism involved, this reaction could be discussed as well in Chapter V on Cope rearrangements.

The few examples of ring enlargement reactions of 2H-azirines and aziridines discussed above demonstrate their vast application in the synthesis of heterocycles. Depending on the conditions, the three-membered rings can be enlarged in one step up to six membered rings, as shown. Presumably, the three atoms of the reactive azirine derivatives can be applied in general to the enlargement of proper functionalized rings of any size.

Cyclopropane and its Derivatives

Carbocyclic three-membered rings can be applied to ring enlargement reactions, too. The thermal expansion of a vinylcyclopropane (**III/39**) to a cyclopentene (**III/40**) ring (Scheme III/7) is a well known ring expansion reaction by two carbon atoms, named vinylcyclopropane rearrangement [28] (the stereochemical specificity was carefully investigated [29] [30]). The reaction has been carried out on many vinylcyclopropanes bearing various substituents in both the ring and the vinyl group. It has been applied to the synthesis of heterocyclic systems, especially to the synthesis of alkaloids. In this modification the vinylgroup is replaced by an immonium residue. Catalytic amounts of anhydrous

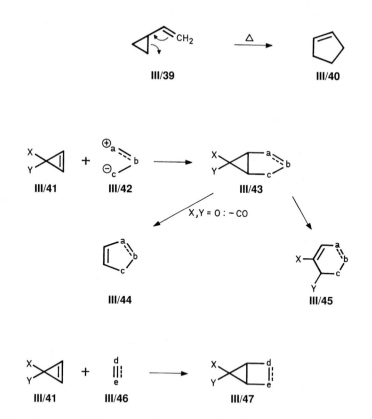

Scheme III/7. The use of the three carbon atoms of cyclopropane as building elements for larger rings.

hydrogenhalide acids and heat induce the rearrangement. Fluoroborate or perchlorate salts fail to catalyze the rearrangement under similar conditions, which proves that the counterion must be nucleophilic. A proposed mechanism is outlined in Scheme III/8 [31]. Two imin residues can be involved, leading to diazepine derivatives e. g. transformation of N,N'-dibenzyliden-trans-1,2-diaminocyclopropane to 2,3-diphenyl-2,3-dihydro-1H-1,4-diazepine at 120° for 1 h in 59 % yield [32]. From the mechanistic point of view the vinylcyclopropane rearrangement is a [1.3] sigmatropic migration of carbon or an internal $[_{\pi}2 + _{\sigma}2]$ cycloaddition reaction [33], Scheme III/7. 2,3-Dihydrofuran occurs as a thermolysis product of 2-vinyloxirane, another example of the heterocyclic version of this isomerization reaction type [34].

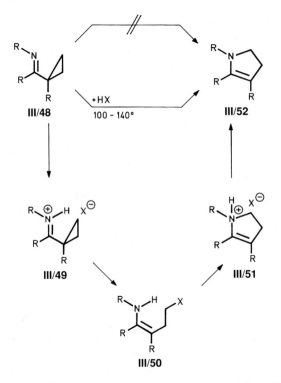

Scheme III/8. The vinylcyclopropane rearrangement as a tool for alkaloid synthesis; a proposed mechanism [31].

HX = HCl, HBr, HI, NH$_4$Cl, NH$_4$Br, NH$_4$I HX ≠ HBF$_4$, HClO$_4$

The following references contain additional examples of this method and corresponding reaction types: Synthesis of 3-fluoroethyl-2-cyclopenten-1-ones [35], preparation of functionalized bicyclo[5.3.0]decane systems and conversion of 1,2-divinylcyclopropanes to functionalized cycloheptanes [36] *e.g.* karahanaenone [37], (±)-*β*-himachalene [38][39]. Metal-catalyzed ring expansions in which cyclopropane (*e.g.* [40] [41] [42] [43] [44]) or aziridines (*e.g.* [45] [46]) and diaziridines (*e.g.* [45]) function as essential moieties are well known reactions.

Although metal-catalyzed ring expansions will not be discussed in this text, these reactions are of great value in synthetic organic chemistry. This restriction is not limited to three-membered ring chemistry only, it is true for all other reaction types. Thus, the following reactions will be excluded: metal-catalyzed insertion of carbon monoxide (carbonylation), *e.g.* [42] [46] [47] [48] or of olefins and alkynes, *e.g.* [49] [50] [51] [52] or of other groups, *e.g.* [53] [54] into cycloalkanes of different ring size. Furthermore, corresponding silicon mediated chemical reactions [50] [51] [52] have had to be eliminated as well.

Cyclopropenes and cyclopropenones can also be used as precursors for expanded ring systems. The intermediates of the ring enlargement reactions are generated by 1,3-dipolar addition or by a [2+2] addition across the cyclopropenyl *π*-bond [55]. The reaction principle is summarized in Scheme III/7. In the addition of a 1,3-dipole (**III/42**) to a cyclopropene, **III/41**, the bicyclic compound **III/43** is formed. Depending on several factors the primary reaction product, **III/43**, may be stable. In cases where **III/43** is formed from cyclopropenone, it may eliminate carbon monoxide to yield the five-membered **III/44**. An alternative reaction possibility for **III/43** is its spontaneous rearrangement to the monocyclic compound, **III/45**. Acylic decomposition products of compound **III/43** are known, too.

Another possibility for the preparation of bicyclic systems such as **III/47** from three-membered rings can be realized by a [2+2] cycloaddition of the cyclopropene, **III/41**, and an unsaturated molecule, **III/46**, such as alkene, alkyne, ketene, ketenimine, ketone, isocyanate, *etc.* A large number of examples of this reaction type have been reviewed recently [55][1].

An example of the newer literature is given in Scheme III/9 [61].
The readily available methyl 3,3-dimethylcyclopropene-1-carboxylate (**III/54**) undergoes [2+2] cycloaddition with enamines *e.g.* the morpholine derivative **III/55** to give 2-aminobicyclo[2.1.0]pentane derivatives, *e.g.* **III/56**. These compounds are transformed into cyclopentane derivatives, *e.g.* methyl 4-hydroxy-2,5,5-trimethyl-1-cyclopentenecarboxylate (**III/57**) by treatment with dilute mineral acids.

---

1) Treatment of cyclopropenones with isonitriles [56] [57] and of triafulvenes with isonitriles [58] gives cyclobutenes. Formations of expanded rings are observed by the reactions of enamines with diphenylcyclopropenones [59] [60].

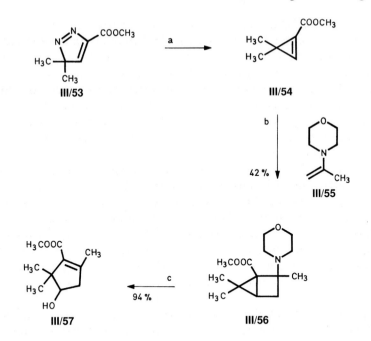

Scheme III/9. Cyclopropene as a three-carbon atom synthon [61].

a) Et$_2$O, hν, −N$_2$ b) Et$_2$O, 20°, 15 h   c) H$_2$SO$_4$, H$_2$O, C$_6$H$_6$, 80°, 4 h.

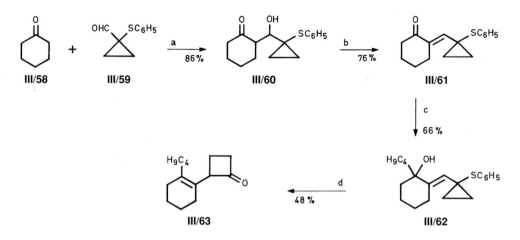

Scheme III/10. Three- to four-membered ring enlargement [62].

a) LiN(i-Pr)$_2$, THF   b) POCl$_3$, hexamethylphosphoramide   c) C$_4$H$_9$MgBr
d) HBF$_4$, (C$_2$H$_5$)$_2$O.

A three- to four-membered ring enlargement reaction has been chosen as a part of the synthetic concept in the rearrangement of compound **III/62** to **III/63**, Scheme III/10. The starting material for the rearrangement was prepared by an aldol reaction of cyclohexanone (**III/58**) and 1-(arylthio)-cyclopropanecarbo-xaldehyde (**III/59**). After dehydration of **III/60**, the resulting α,β-unsaturated ketone **III/61** was treated with a nucleophile to give the tertiary alcohol, **III/62**. The latter is vinylogous to cyclopropylcarbinols, which are known to rearrange to cyclobutanones. The reaction is acid catalyzed (48 % fluoroboric acid), and the yields are moderate [62].

Cyclopropane derivatives of type **III/64** (Scheme III/11) have been shown to be useful starting materials for a smooth transformation to furanones [63] and thiophenes [64]. The aldol reaction of **III/64** and a ketone or aldehyde yielded **III/65**, which forms, on desilylation, an ester diol (by a retro aldol reaction).

In this reaction only the cyclopropane ring is opened to form **III/66**. This compound is in equilibrium with the γ-lactol **III/67**, which, in case of R=H, was transformed oxidatively (pyridinium chlorochromate) to the furanone **III/68**. The synthesis of thiophene derivatives in a similar reaction is shown in Scheme III/11, also.

In several synthetic studies, cyclopropane derivatives were used as synthones or building elements for ring enlargement steps, *e.g.* reaction of enamines with cyclopropenone [65], synthesis of 2,3-dihydro-1,4-diazepine by thermal iso-merization of 1,2-diamino-cyclopropanes [32] [66], and preparation of 3-amino-fulvenes from methylencyclopropenes with alkynamines [67].

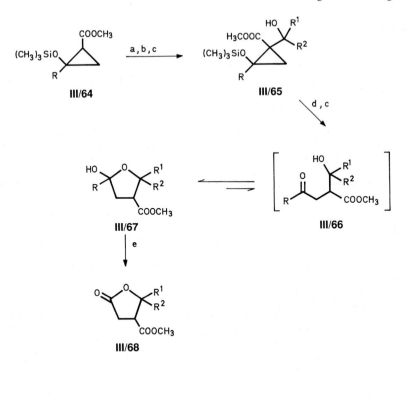

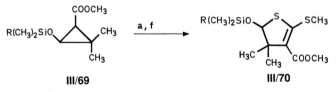

Scheme III/11. Formation of furanone and thiophene derivatives by rearrangement of methyl 2-siloxycyclopropanecarboxylates [61] [62] [63] [64].

a) Lithium diisopropylamide   b) R^1R^2CO c) $H_3O^{\oplus}$
d) Bu_4NF e) R=H, pyridinium chlorochromate, CH_2Cl_2 f) $CS_2 - CH_3I$.

References

[1] E. Vogel, Angew.Chem. **72**, 4 (1960).
[2] J. M. Conia, M. J. Robson, Angew.Chem. **87**, 505 (1975), Angew.Chem.Int.Ed. Engl. **14**, 473 (1975).
[3] H. Heimgartner, Israel J.Chem. **27**, 3 (1986).
[4] H. Heimgartner, Chimia **33**, 111 (1979).
[5] B. Scholl, J. H. Bieri, H. Heimgartner, Helv.Chim.Acta **61**, 3050 (1978).
[6] S. M. Ametamey, R. Hollenstein, H. Heimgartner, Helv.Chim.Acta **71**, 521 (1988).
[7] S. Chaloupka, P. Vittorelli, H. Heimgartner, H. Schmid, H. Link, K. Bernauer, W. E. Oberhänsli, Helv.Chim.Acta **60**, 2476 (1977).
[8] M. Schläpfer-Dähler, R. Prewo, J. H. Bieri, G. Germain, H. Heimgartner, Chimia **42**, 25 (1988).
[9] B. Hostettler, J. P. Obrecht, R. Prewo, J. H. Bieri, H. Heimgartner, Helv.Chim.Acta **69**, 298 (1986).
[10] M. Schläpfer-Dähler, R. Prewo, J. H. Bieri, H. Heimgartner, Heterocycles **22**, 1667 (1984).
[11] O. C. Dermer, G. E. Ham, "Ethyleneimine and other aziridines", Academic Press, New York, 1969.
[12] S. Chaloupka, H. Heimgartner, Chimia **32**, 468 (1978).
[13] S. Chaloupka, H. Heimgartner, Helv.Chim.Acta **62**, 86 (1979).
[14] H. Heimgartner, Wiss.Z.Karl-Marx-Univ. Leipzig, Math.-Naturwiss. R. **32**, 365 (1983).
[15] H. Heimgartner, Chimia **34**, 333 (1980).
[16] C.W. Bird, G.W. H. Cheeseman, "Synthesis of Five-membered Rings with One Hetero-atom" in "Comprehensive Heterocyclic Chemistry", A. R. Katritzky, C.W. Rees (eds.) Vol. 4 (1984), Pergamon Press, Oxford p. 89–153.
[17] A. J. Boulton, A. McKillop, "Synthesis of Six-membered Rings" in "Comprehensive Heterocyclic Chemistry" A. R. Katritzky, C.W. Rees (eds.) Vol. 2 (1984), Pergamon Press, Oxford p. 67–98.
[18] M. E. Freed, A. R. Day, J.Org.Chem. **25**, 2108 (1960).
[19] G. DeStevens, M. Sklar, J.Org.Chem. **28**, 3210 (1963).
[20] H. Stamm, Chem.Ber. **99**, 2556 (1966).
[21] J. Lehmann, H. Wamhoff, Synthesis **1973**, 546.
[22] F. Jakob, P. Schlack, Chem.Ber. **96**, 88 (1963).
[23] D. C. Rees, J.Heterocycl.Chem. **24**, 1297 (1987).
[24] A. Padwa, D. Dean, T. Oine, J.Am.Chem.Soc. **97**, 2822 (1975).
[25] E. Vedejs, J.W. Grissom, J. K. Preston, J.Org.Chem. **52**, 3488 (1987).
[26] J. A. Deyrup, S. C. Clough, J.Am.Chem.Soc. **91**, 4590 (1969).
[27] N. Manisse, J. Chuche, J.Am.Chem.Soc. **99**, 1272 (1977).
[28] C. D. Gutsche, D. Redmore, "Carbocyclic Ring Expansion Reactions", Academic Press, New York 1968.
[29] G. D. Andrews, J. E. Baldwin, J.Am.Chem.Soc. **98**, 6705 (1976).
[30] G. D. Andrews, J. E. Baldwin, J.Am.Chem.Soc. **98**, 6706 (1976).
[31] R.V. Stevens, Acc.Chem.Res. **10**, 193 (1977).
[32] H. A. Staab, F. Vögtle, Chem.Ber. **98**, 2701 (1965).
[33] J. March, "Advanced Organic Chemistry", 3. Ed., John Wiley & Sons, New York 1985.
[34] R. J. Crawford, S. B. Lutener, R. D. Cockcroft, Can.J.Chem. **54**, 3364 (1976).
[35] M. Shimizu, H. Yoshioka, Tetrahedron Lett. **28**, 3119 (1987).
[36] P. A. Wender, M. A. Eissenstat, M. P. Filosa, J.Am.Chem.Soc. **101**, 2196 (1979).
[37] P. A. Wender, M. P. Filosa, J.Org.Chem. **41**, 3490 (1976).
[38] E. Piers, E. H. Ruediger, J.Chem.Soc., Chem.Commun. **1979**, 166.
[39] E. Piers, I. Nagakura, H. E. Morton, J.Org.Chem. **43**, 3630 (1978).
[40] G. Albelo, G. Wiger, M. F. Rettig, J.Am.Chem.Soc. **97**, 4510 (1975).

[41] H. R. Beer, P. Bigler, W. v.Philipsborn, A. Salzer, Inorg.Chim.Acta **53**, L49 (1981).
[42] S. L. Buchwald, B. T. Watson, J. C. Huffman, J.Am.Chem.Soc. **109**, 2544 (1987).
[43] J. P. Marino, L. J. Browne, Tetrahedron Lett. **1976**, 3245.
[44] W. A. Donaldson, B. S. Taylor, Tetrahedron Lett. **26**, 4163 (1985).
[45] D. Roberto, H. Alper, J.Chem.Soc., Chem.Commun. **1987**, 212.
[46] S. Calet, F. Urso, H. Alper, J.Am.Chem.Soc. **111,** 931 (1989).
[47] H. C. Brown, A. S. Phadke, M. V. Rangaishenvi, J.Am.Chem.Soc. **110**, 6264 (1988).
[48] D. Roberto, H. Alper, J.Am.Chem.Soc. **111**, 7539 (1989).
[49] S. Iyer, L. S. Liebeskind, J.Am.Chem.Soc. **109**, 2759 (1987).
[50] H. Sakurai, Y. Kamiyama, Y. Nakadaira, J.Am.Chem.Soc. **97**, 931 (1975).
[51] H. Sakurai, T. Imai, Chem.Lett. **1975**, 891.
[52] H. Sakurai, T. Kobayashi, Y. Nakadaira, J. Organomet. Chem. **162**, C43 (1978).
[53] J. E. Baldwin, R. M. Adlington, T. W. Kang, E. Lee, C. J. Schofield, J.Chem.Soc., Chem.Commun. **1987**, 104.
[54] D. P. Klein, J. C. Hayes, R. G. Bergman, J.Am.Chem.Soc. **110**, 3704 (1988).
[55] M. L. Deem, Synthesis **1982**, 701.
[56] N. Obata, T. Takizawa, Tetrahedron Lett. **1970**, 2231.
[57] J. S. Chickos, J.Org.Chem. **38**, 3643 (1973).
[58] T. Eicher, U. Stapperfenne, Synthesis **1987**, 619.
[59] J. Ciabattoni, G. A. Berchtold, J.Am.Chem.Soc. **87**, 1404 (1965).
[60] J. Ciabattoni, G. A. Berchtold, J.Org.Chem. **31**, 1336 (1966).
[61] M. Franck-Neumann, M. Miesch, H. Kempf, Synthesis **1989**, 820.
[62] B. M. Trost, L. N. Jungheim, J.Am.Chem.Soc. **102**, 7910 (1980).
[63] C. Brückner, H.-U. Reissig, J.Org.Chem. **53**, 2440 (1988).
[64] C. Brückner, H.-U. Reissig, Liebigs Ann.Chem. **1988**, 465.
[65] M. A. Steinfels, A. S. Dreiding, Helv.Chim.Acta **55**, 702 (1972).
[66] H. Quast, J. Stawitz, Tetrahedron Lett. **1977**, 2709.
[67] T. Eicher, T. Pfister, Tetrahedron Lett. **1972**, 3969.

IV. Ring Expansion from Four-membered Rings or via Four-membered Intermediates

IV.1. Ring Expansion from Four-membered Rings

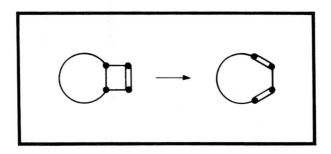

In many reactions, cyclobutane derivatives are intermediates which lead to ring enlarged compounds. Within the scope of this book it will only be possible to discuss reactions which may be of common preparative interest.

Quite recently a general method leading to a large variety of substituted quinones has been developed, *via* thermolysis and subsequent oxidation (air or Ce(IV)salts or Ag₂O) of 4-hydroxy-cyclobutenones substituted in 4-position [1] [2]. The appropriate derivatives of cyclobutenones are prepared from the corresponding cyclobutenediones (Scheme IV/1). The symmetrical **IV/1** is alkylated by an aryl or heteroaryl or alkynyl lithium (or Grignard) reagent. In case of an unsymmetrically substituted cyclobutenedion, high regioselectivity is observed. 3-Methoxy-4-methylcyclobutenedione, for example, reacts with high regioselectivity to 2-hydroxy-3-methoxy-4-methyl-cyclobutenone substituted at position 2 [1].

On heating a xylene solution of compounds such as **IV/2** clean transformations will occur within the range of 20 min to 4 h to produce the corresponding quinones, after oxidation. Judging from experiments on the thermolysis of cyclobutenes and cyclobutenones (see below), the reaction probably occurs through a vinyl ketene. These transformations are thought to be dictated by a favored conrotatory ring opening of **IV/2**, so that the electrondonating substituents (OH, OR) rotate outward. Thus, the ketene **IV/3**, will have the precise

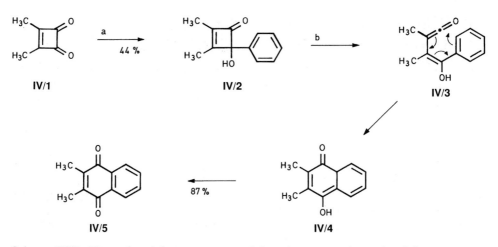

Scheme IV/1. Thermal cyclobutenone → naphthoquinone transformation [1].
a) 1. C$_6$H$_5$Li, THF,-78° 2. NH$_4$Cl   b) 1. 160°, 5 min 2. air.

configuration for a direct interaction of its electrophylic site, with the proximal
aryl or heteroaryl or alkyne group. Intermediate **IV/4** is oxidized to give the
quinone **IV/5**. The conversion of the cyclobutenone allyl ether **IV/6** to the
benzoquinone **IV/9** (76 % yield) is a fascinating example, Scheme IV/2 [2]. This
rearrangement is predetermined to involve the ketene **IV/7**, which cyclizes to
the zwitterionic species **IV/8**. Subsequently, intramolecular electrophilic attack
on the allylic bond, and C-O bond cleavage leads to product **IV/9** [2].

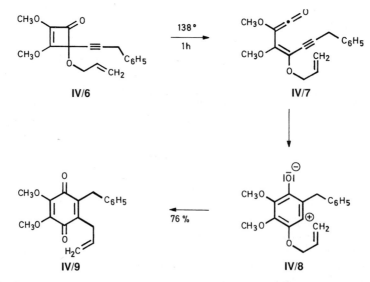

Scheme IV/2. Thermal cyclobutenone → benzoquinone transformation involving an allyl
group migration [2].

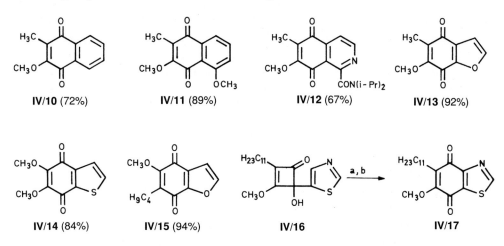

IV/10 (72%)   IV/11 (89%)   IV/12 (67%)   IV/13 (92%)

IV/14 (84%)   IV/15 (94%)   IV/16   a, b   IV/17

Scheme IV/3. Some examples of quinones from 4-aryl-(or heteroaryl)-4-hydroxybutenones [1] [2] and [3] [4].

a) 138°   b) Ag$_2$O.

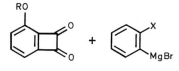

IV/18   IV/19

a

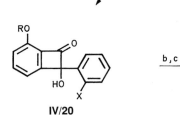

IV/20   IV/21

major isomer   major isomer

·) R = CH$_3$, X = H: 81 %   81 %
  regioisomere ratio: 6.6 : 1
·) R = CH$_3$, X = OCH$_3$: 70 %   96 %
  regioisomere ratio: 3.5 : 1
·) R = (CH$_3$)$_2$tBuSi, X = H: 80 %   76 % (R = X = H, after F$^\ominus$ treatment)
  regioisomere ratio: > 20 : 1
·) R = (CH$_3$)$_2$tBuSi, X = OCH$_3$: 67 %   89 % (R = H, X = CH$_3$, after F$^\ominus$ treatment
  regioisomere ratio: > 20 : 1

Scheme IV/3 gives the yields of the quinones, **IV/10** to **IV/15** synthesized by this method. Regiospecificity in the formation of anthraquinones is demonstrated in Scheme IV/4 [1].

A similar reaction is observed when 4-alkynyl-2,3-dimethoxy-4-(trimethyl-siloxy)cyclobutenones are thermolyzed in xylene under reflux. Five- as well as six-membered ring compounds result [5]: The cyclobutenone **IV/22** (Scheme IV/5) is thought to be in equilibrium with the ketene **IV/23**. The ring closure of **IV/23** to **IV/24** and **IV/26** is influenced by the substituent R. Electron-with-drawing groups (such as $COOC_2H_5$) favor the formation of cyclopentenedione **IV/27** and electon-releasing groups encourage the formation of quinones **IV/25**, Scheme IV/5.

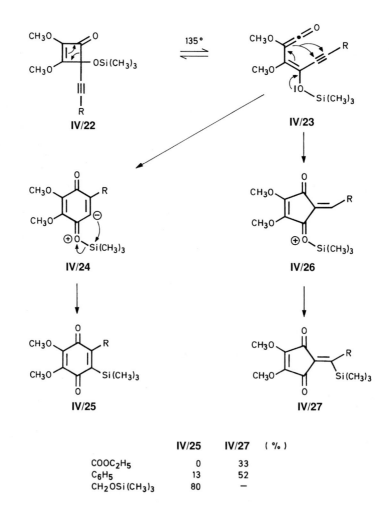

	IV/25	IV/27	(%)
$COOC_2H_5$	0	33	
C_6H_5	13	52	
$CH_2OSi(CH_3)_3$	80	–	

Scheme IV/5. Thermolyses of cyclobutenones and formation of six- or five-membered ring compounds [5].

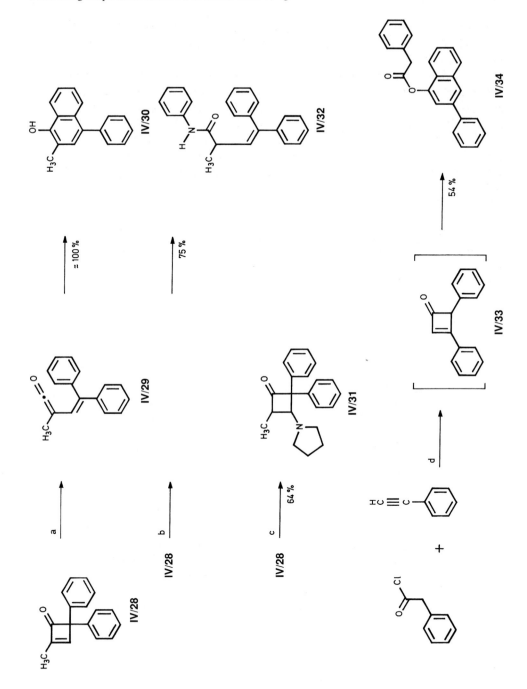

Scheme IV/6. Thermal rearrangement of cyclobutenones to phenols.

a) Cyclohexane, 1 h reflux   b) aniline 100°   c) pyrrolidine 80°
d) 170–180°, 15 h.

The thermal rearrangement of cyclobutenones to naphthols represents a ring enlargement reaction or, depending on the reaction conditions, a ring formation reaction. While the latter type is not a subject of this review, the former is.

The cyclobutenone **IV/28**, in boiling cyclohexane, Scheme IV/6, is transformed to the α-naphthol **IV/30** in nearly quantitative yield [6]. If the reaction is carried out in aniline at 100°, the anilide **IV/32** is obtained [7]. Compound **IV/32** is formed by a nucleophilic reaction of aniline on the vinylketene intermediate **IV/29** preventing ring closure. A strong nucleophile (*e.g.* pyrrolidine) can add to the endocyclic double bond of the cyclobutenone **IV/28** to yield the cyclobutanone **IV/31**, again preventing ring expansion [7] [8]. When phenylacetyl chloride was treated with phenylethyne, neat, at 170–180° for 15 h [9][1)] a cyclobutenone, probably **IV/33**, was an intermediate in the formation of **IV/34**. – Another way to prepare phenols is given in Scheme IV/7. Thermolysis of cyclobutenones in the presence of alkynes leads in high yield to substituted phenols [10]. The reaction mechanism may involve the formation of the four-membered intermediate **IV/37** or, as an alternative, the six-membered ketone **IV/37a**, which offers a shorter route.

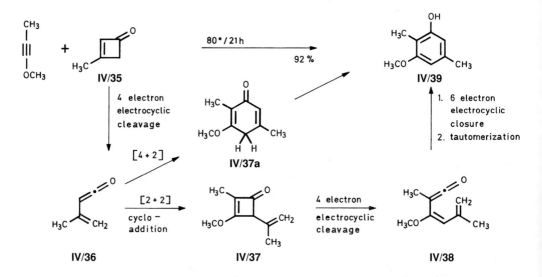

Scheme IV/7. Similar phenol forming reactions have been published in ref. [11] [12] [13] [14].

---

1) It is possible that the first intermediate formed was a ketene (*e.g.* **IV/29**) since, if the reaction was carried out with the same ketene as starting material, the corresponding products (*e.g.* **IV/30**) were isolated [15] [16].

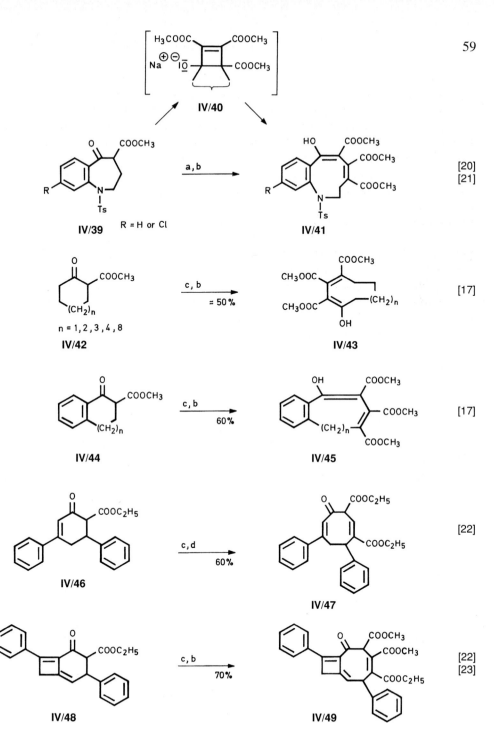

Scheme IV/8. (n+2) Conversion of cyclic compounds by acetylene derivatives.
a) CH_3OH/CH_3ONa, Δ 2 h   b) $H_3COOCC \equiv CCOOCH_3$ c) NaH, toluene
d) $HC \equiv CCOOC_2H_5$.

Cyclic active-methylene compounds react in Michael reactions not only with α,β-unsaturated acceptors, but also with acetylene dicarboxylates, propiolates, and their derivatives (Scheme IV/8). If, for instance, acetylene carboxylic esters are added to a variety of cyclic-active methylene compounds, ring-expanded products generally are formed directly. In other cases, it is possible to isolate the intermediate cyclobutene derivatives, which can be converted to the ring enlarged compounds in a second reaction step.

Cyclic β-ketoesters have been treated with strong base (NaH/toluene; CH₃ONa/CH₃OH) and afterwards with acetylenecarboxylates or ethyl propiolates. From the reaction mixture the corresponding cycloalkenes, enlarged by two carbon atoms, can be isolated in 50–70 % yield. In the transformations of β-ketoesters, given in Scheme IV/8, no intermediate could be detected. Therefore and because of the strong basic conditions used, the concept of a concerted reaction must be rejected in these cases, and a simple polar concept is preferred [17].

The dimethylsulfoniumylide **IV/50**, prepared from the corresponding fluoroborate by the action of sodium hydride, has been treated with dimethyl acetylenedicarboxylate in dimethyl sulfoxide to form the eight-membered **IV/52** in quantitative yield [18] [19], Scheme IV/9.

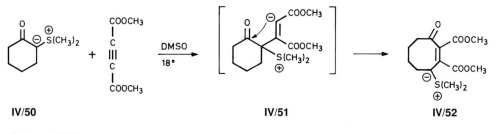

IV/50   **IV/51**   **IV/52**

Scheme IV/9.

In other examples acetylenecarboxylic, if added to a variety of enamines of cycloalkanones, give the ring-expanded product directly in most cases. The conditions are very mild. In general, the reactions of propiolates with a number of enamines involve initial [2+2] cycloaddition to form cyclobutene derivatives. These cyclobutenes rearrange, quite often spontaneously, sometimes producing ring enlargement by two carbon atoms. However, the conditions, yields, numbers, and types of by-products vary strongly depending on the ring size of the enamine and the cycloalkanones [24], Scheme IV/10.

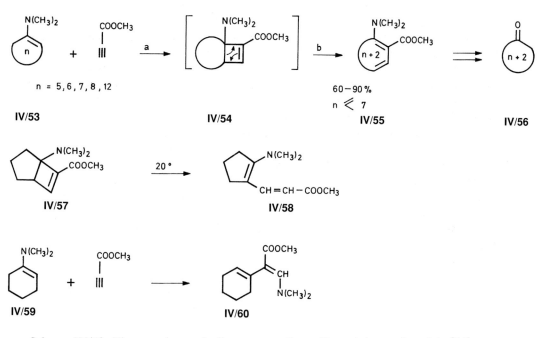

Scheme IV/10. The enamine cycloalkanone reaction with acetylenecarboxylate [24].

a) [2+2] Cycloaddition   b) cycloreversion.

It is remarkable that the bicyclic intermediate **IV/54** could be isolated, in all the reactions which have been investigated. The five-membered enamine derivative **IV/57** when kept at room temperature, isomerizes to the monocyclic **IV/58**. The treatment of 1-dimethylamino-cyclohexene (**IV/59**) with methyl propiolate gave no bicyclic analogue of type **IV/54** and no monocyclic analogue of **IV/55**. Instead, only the rearrangement product, **IV/60**, was observed. A possible rationalization is given in ref. [24]. Reactions of enamines with acetylenic esters have been used to synthesize *bis*-homosteroids from steroids [25], an analogue of the antileukemic constituent of *Steganotaenia araliacea* Hochst., steganone [26], and benzazepines from indoles [27] [28] [29] [30]. They have also been applied to enlarge cyclic ketones [24] [31] [32], to investigate the reaction behavior of cyclododecanone [33], benzoxocines and benzazocines from quinolines and coumarines [34], to synthesize (±)-muscone from cyclododecanone *via* two- and one-carbon atom enlargement [24][35], and to prepare (±)-muscone from cyclotridecanone [36].

The reaction of five-membered heterocycles with acetylenecarboxylates under conditions of cycloaddition occasionally yielded bicyclic intermediates which rearranged to the ring expanded products on heating; compare examples (1) and (2) in Scheme IV/11, and Chapter III, Scheme III/4. If more polar starting materials (Scheme IV/11, entries 3 and 4) are used, it is not possible to

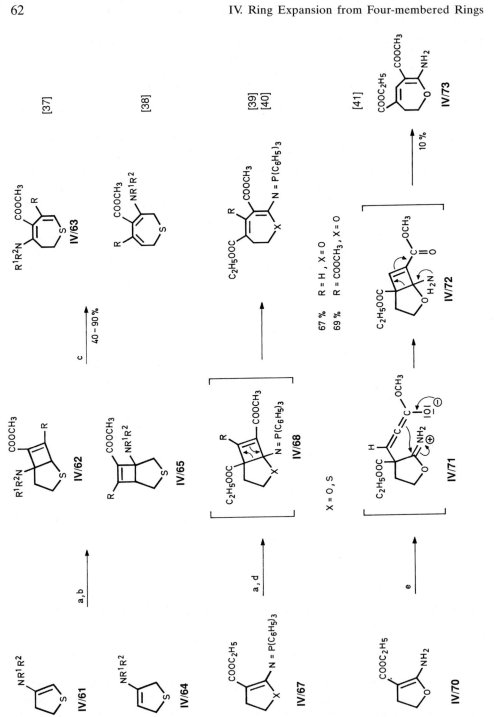

Scheme IV/11. Reactions of heterocyclic systems with acetylenecarboxylates.

R = H, COOCH₃; R¹,R² = alkyl, X = O, S
a) RC≡C-COOCH₃   b) (C₂H₅)₂O, 20°, 16 h   c) dioxane, 100°, 2 h
d) CH₃CN, 2 h reflux   e) HC≡C-COOMe, toluene, 20°.

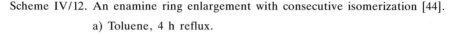

detect the corresponding bicyclic butene intermediates (**IV/68** or **IV/72**), even if the reaction is carried out at room temperature [41]. For similar enamine acetylenic reactions compare ref. [42] [43]. However such bicyclic butenes can be proved to be intermediates by indirect methods, see the example shown in Scheme IV/12. Careful treatment of the seven-membered enamine with diethyl acetylenedicarboxylate gave the nine-membered ring compound **IV/75** with a (Z/Z) configuration in the resulting butadiene. In the following thermal reaction, the isomerization to the (Z,E) configured isomer **IV/76** can be observed [44] [45], Scheme IV/12. The formation of compound **IV/75** can be easily explained by a conrotatory opening of a bicyclic butene intermediate.

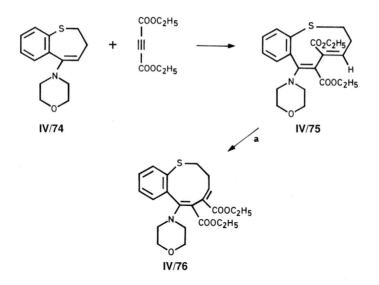

IV/74   **IV/75**

a

IV/76

Scheme IV/12. An enamine ring enlargement with consecutive isomerization [44].

a) Toluene, 4 h reflux.

Cyclobutenes properly substituted by a cyclic 1,2 or 1,3 substituent undergo thermal cycloreversion to give ring-enlarged products. The energy of the opening [46] of the bicyclic compound is reduced if a double bond is present which will be in conjugation with the new one. Thus, compound **IV/77** can be pyrolyzed to form the cycloheptatrienone **IV/78**. The corresponding reaction is not

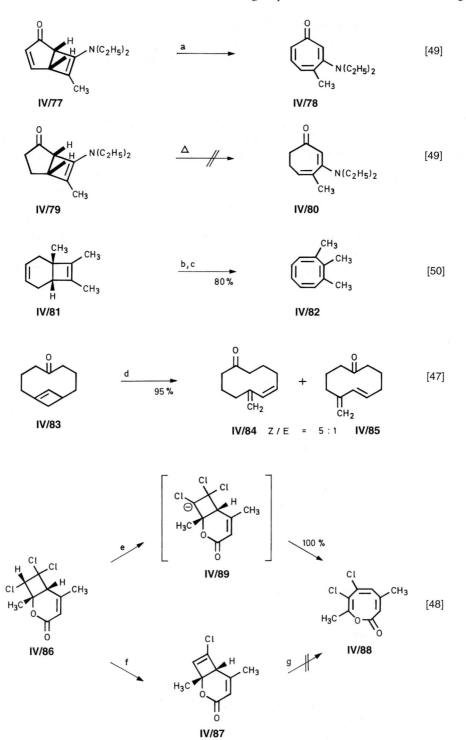

Scheme IV/13. Further examples of cyclobutene rearrangements.

a) 400°, flash pyrolysis   b) Br$_2$ c) 60°   d) toluene, 180°
e) (C$_2$H$_5$)$_3$N, reflux   f) Zn (for dechlorination)   g) C$_2$H$_5$OH, reflux.

◄──

possible for the dihydro derivative **IV/79** (→ **IV/80**), see Scheme IV/13. The two step transformation, **IV/81** → **IV/82**, is facilitated by double bonds introduced by a bromination/dehydrobromination reaction. The synthetically useful conversion of bicyclic **IV/83**[2)] to the two ten-membered isomers, **IV/84** and **IV/85**, in a 5:1 (*Z* to *E*) ratio, which was part of a germacrane synthesis [47], is another example of a thermal cyclobutene cyclorevision.

The reaction of the photoadduct, **IV/86** (4,6-dimethyl-2-pyrone and trichloroethylene) with triethylamine in ethanol yielded the dehydrochlorinated product, **IV/88**, Scheme IV/13. Reaction of **IV/86** with Zn dust in refluxing acetonitrile gave **IV/87**[3)] which proved to be resistant to boiling ethanol [48], thus contrasting with the examples given above.

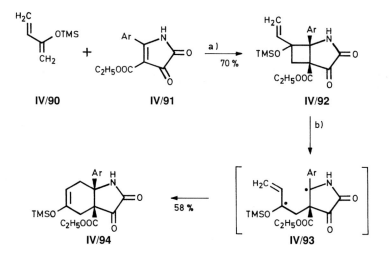

Scheme IV/14. A specific ring enlargement reaction [51].

Ar = C$_6$H$_5$.
a) Dimethoxyethane, high-pressure mercury lamp, 0°, 30 min
b) toluene, 3 h, reflux.

---

2) One of the structure proofs of **IV/83** was the fully decoupled ^{13}C-NMR spectrum of its dihydro-derivative, which displayed only six carbon signals.
3) When heated in refluxing toluene for 80 h, **IV/87** was converted to 6-chloro-2-hydroxy-4-methylacetophenone [48].

A homolytic cleavage in a four-membered ring, followed by an allylic migration, can enlarge a ring by two members. The four-membered ring compound of type **IV/92** was easily prepared by photocycloaddition of 2-trimethylsilyloxybuta-diene (**IV/90**) and the pyrrolidine dione **IV/91**. The enlargement reaction (**IV/92** → **IV/94**) proceeds because radical stabilizing functional groups are present [51], Scheme IV/14.

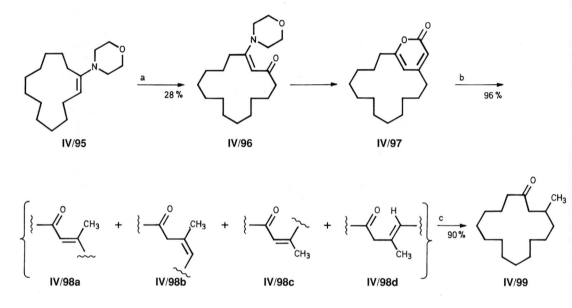

Scheme IV/15. Application of the enamine-ketene ring enlargement for the synthesis of (±)-muscone (**IV/99**) [53].

a) 2 CH₂=C=O, CHCl₃   b) KOH, H₂O, C₂H₅OH, reflux
c) H₂-Pd/C.

An additional example is a two-step approach to the preparation of 1,5-cyclo-decadiene. By a photochemical cycloaddition of a substituted cyclobutene and 2-cyclohexenone, a strained tricyclo[4.4.0.0²,⁵]decane system is generated. Thermolysis of the tricycle gave 1,5-cyclodecadiene [52].

Another ketene reaction, the enamine ketene reaction, can also be used for ring enlargement. An application of this reaction is shown in Scheme IV/15, in which the enamine of morpholine and cyclotridecanone **IV/95** has been transformed to (±)-muscone (**IV/99**) [53] [54] [55]. Two moles of ketene – generated *in situ* from acetylchloride and triethylamine or introduced as gaseous ketene – were condensed step-wise with the enamine **IV/95** to form the α-pyrone, **IV/97**. The yield is low (10–28 %). Base catalyzed hydrolysis of **IV/97** gave a mixture of the four isomers, **IV/98a** to **IV/98d**, yielding, after catalytic hydrogenation, (±)-muscone (**IV/99**) [53].

IV.2. Benzocyclobutene Derivatives as Intermediates

Several benzannelated compounds have been isolated from natural sources and synthesized. However, only one general method, as described below, has been devised for their preparation.

Ketone enolates prepared from ketones and sodium amide generate arynes from halobenzenes. If a cyclic ketone enolate is used in excess, both components, the aryne and the ketone enolate, can react with each other, forming a ring enlarged benzannelated cycloalkanone. The mechanism of this reaction, called arynic condensation, is given in Scheme IV/16 [56] [57]. Reaction of the aryne, **IV/102**, with the sodium enolate, **IV/103**, gives first the salt **IV/104**. This salt reacts to give a tricyclic anion, **IV/105**, which contains a four-membered ring. In **IV/105** the alcoholate-oxygen is placed in benzylic position. The ring enlarged bicyclic anion, **IV/106**, is an isomeric structure to **IV/105**. By proton transfer and finally by protonation, the latter yields the benzannelated cycloalkanone, **IV/109**, as the main product. Beside compound **IV/109**, two other products have been isolated in this kind of reaction: 2-phenylcycloalkanone, **IV/107**, and the tricycle, **IV/108**, by protonation of **IV/104** and **IV/105**. The third product is the ring opened primary amide, **IV/110**, derived from **IV/107** or **IV/109** by nucleophilic attack of ammonia. A number of experimental and structural features influence the total yields and the ratios of the individual products [57].

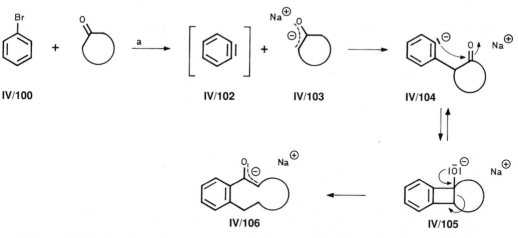

Scheme IV/16. Arynic condensation of ketone enolates [56] [57].

a) NaNH$_2$, 1,2-dimethoxyethane, 45°, 6 h, molar ratio **IV/103:IV/100** ≈ 2:1.

The yield of the tricyclic **IV/108** increases in the presence of Li$^+$ instead of Na$^+$ while that of the 2-phenylketone, **IV/107**, decreases (see Scheme IV/17 for the cyclohexanone reaction). Because of its better oxygen complexing ability the anion **IV/105** is more stabilized by lithium than by the sodium cation [57]. Compound **IV/108** is not observed if NaI is added to the reaction mixture just before hydrolysis. The only products isolated under these conditions are **IV/107** and **IV/109** (with cycloheptanone), the tricycle is missing. A competition for Na⁺ between the alcoholate oxygen and iodide was suggested. Under these circumstances the anion **IV/105** behaves as a benzocyclobutenol [58] to form **IV/104** and **IV/106** and, after hydrolysis, **IV/107** and **IV/109** [57].

Compounds of type **IV/108** may be opened by reaction with a base in aprotic solvent to give the ring enlarged product in very good yield [56] [59]. This also demonstrates that **IV/104** and **IV/105** are in equilibrium. Prolongation of the reaction time only leads to higher yields of **IV/107** and ring opened **IV/110** and with a lower amount of **IV/108**.

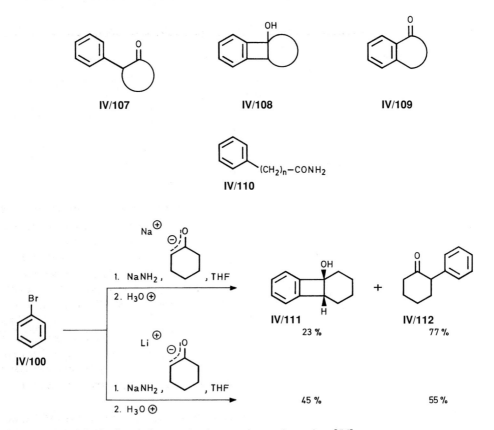

Scheme IV/17. Cation influence in the arynic condensation [56].

The reaction can be carried out in two solvents, tetrahydrofuran and 1,2-dimethoxyethane (DME). DME allows condensations at lower temperatures but also favors the formation of benzocyclobutenoates **IV/108**. The non-nucleophilic complex base, NaNH$_2$-t-BuONa, used with low reactive ketones, promotes the formation of ketonic compounds **IV/107** and **IV/109**. The yields in some examples are given below [56] [57] [60].

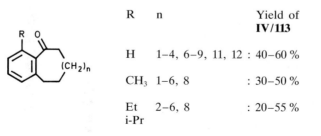

R	n	Yield of **IV/113**
H	1–4, 6–9, 11, 12	: 40–60 %
CH$_3$	1–6, 8	: 30–50 %
Et	2–6, 8	: 20–55 %
i-Pr		

The arynic condensation, yields of ring enlargement products **IV/113**. For additional informations see ref. [61] [62].

The arynic condensation of 1,2-diketone monoacetal enolates with aryne (from bromobenzene and sodium amide) lead to **IV/114**. Depending on the reaction conditions, compound **IV/114** can be transformed to two different enlargement products, **IV/115** and **IV/117**. The mechanisms involved in these transformations are obvious; [63] [64], Scheme IV/18.

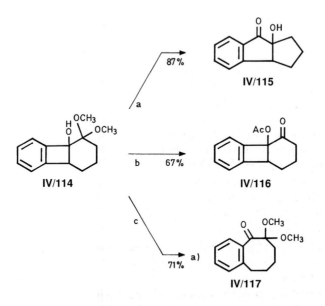

Scheme IV/18. Reactions of a cyclobutenol [63]. The yields of higher homologues of **IV/117** are in the range of 80–96 %.

a) CH$_3$OH, H$_3$O$^\oplus$
b) 1. Ac$_2$O, CH$_2$Cl$_2$, 4-dimethylaminopyridine 2. dilute HCl
c) NaNH$_2$, hexamethylphosphoramide.

An enlargement reaction somehow related to the arynic condensation mentioned above is the transformation shown in Scheme IV/19 [65].

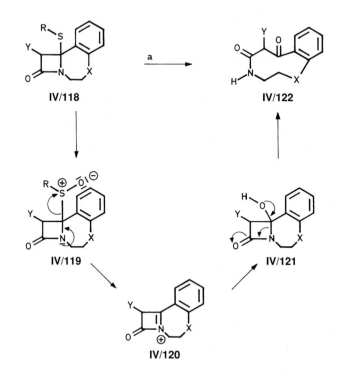

Scheme IV/19. Ring enlargement reaction of β-lactam type [65].

R = Alkyl, X = S or O, Y = OMe or N₃
a) NaIO₄, H₂O, i-PrOH.

Oxidation of the β-lactam **IV/118** with $NaIO_4$ in a water-isopropanol solution leads to the benzannelated nine-membered lactam **IV/122**. Probably the sulfide is oxidized to a sulfoxide, **IV/119**, in the first reaction step. As the better leaving group, the sulfoxide is ejected by the β-lactam nitrogen atom. The immonium double bond is then hydrolyzed to a ketone and a lactam in a nine-membered ring.

References

[1] L. S. Liebeskind, S. Iyer, C. F. Jewell, J.Org.Chem. **51**, 3065 (1986).
[2] S.T. Perri, L. D. Foland, O. H.W. Decker, H.W. Moore, J.Org.Chem. **51**, 3067 (1986).
[3] D. L. Selwood, K. S. Jandu, Trop.Med.Parasitol. **39**, 81 (1988).
[4] D. L. Selwood, K. S. Jandu, Heterocycles **27**, 1191 (1988).
[5] J. O. Karlsson, N.V. Nguyen, L. D. Foland, H.W. Moore, J.Am.Chem.Soc. **107**, 3392 (1985).
[6] H. Mayr, Angew.Chem. **87**, 491 (1975), Angew.Chem.Int.Ed.Engl. **14**, 500 (1975).
[7] R. Huisgen, H. Mayr, J.Chem.Soc., Chem.Commun. **1976**, 55.
[8] H. Mayr, R. Huisgen, J.Chem.Soc., Chem.Commun. **1976**, 57.
[9] C. Kipping, H. Schiefer, K. Schönfelder, J.prakt.Chem. **315**, 887 (1973).
[10] R. L. Danheiser, S. K. Gee, J.Org.Chem. **49**, 1672 (1984).
[11] E.W. Neuse, B. R. Green, Liebigs Ann.Chem. **1974**, 1534.
[12] J. Nieuwenhuis, J. F. Arens, Rec.Trav.Chim.Pays-Bas **77**, 1153 (1958).
[13] Z. Zubovics, H. Wittmann, Liebigs Ann.Chem. **765**, 15 (1972).
[14] H. Wittmann, V. Illi, H. Sterk, E. Ziegler, Monatsh.Chem. **99**, 1982 (1968).
[15] L. I. Smith, H. H. Hoehn, J.Am.Chem.Soc. **61**, 2619 (1939).
[16] L. I. Smith, H. H. Hoehn, J.Am.Chem.Soc. **63**, 1181 (1941).
[17] A. J. Frew, G. R. Proctor, J.Chem.Soc., Perkin Trans. I **1980**, 1245.
[18] T. Mukaiyama, M. Higo, Tetrahedron Lett. **1970**, 5297.
[19] M. Higo, T. Sakashita, M. Toyoda, T. Mukaiyama, Bull.Chem.Soc.Jpn. **45**, 250 (1972).
[20] A. J. Frew, G. R. Proctor, J.V. Silverton, J.Chem.Soc., Perkin Trans. I **1980**, 1251.
[21] M. Lennon, A. McLean, I. McWatt, G. R. Proctor, J.Chem.Soc., Perkin Trans. I **1974**, 1828.
[22] M. M. Abou-Elzahab, S. N. Ayyad, M.T. Zimaity, Z. Naturforschung **41b**, 363 (1986).
[23] E. M. Afsah, M. M. A. Elzahab, M.T. Zimaity, G. R. Proctor, Z. Naturforschung **39b**, 1286 (1984).
[24] K. C. Brannock, R. D. Burpitt, V.W. Goodlett, J. G. Thweatt, J.Org.Chem. **29**, 818 (1964).
[25] A. K. Bose, G. Mina, M. S. Manhas, E. Rzucidlo, Tetrahedron Lett. **1963**, 1467.
[26] D. Becker, L. R. Hughes, R. A. Raphael, J.Chem.Soc., Chem.Commun. **1974**, 430.
[27] R. M. Acheson, J. N. Bridson, T. S. Cameron, J.Chem.Soc., Perkin Trans. I **1972**, 968.
[28] H. Plieninger, D. Wild, Chem.Ber. **99**, 3070 (1966).
[29] T. Oishi, S. Murakami, Y. Ban, Chem.Pharm.Bull.Jpn. **20**, 1740 (1972).
[30] M.-S. Lin, V. Snieckus, J.Org.Chem. **36**, 645 (1971).
[31] G. A. Berchtold, G. F. Uhlig, J.Org.Chem. **28**, 1459 (1963).
[32] C. F. Huebner, L. Dorfman, M. M. Robinson, E. Donoghue, W. G. Pierson, P. Strachan, J.Org.Chem. **28**, 3134 (1963).
[33] M. A. Steinfels, A. S. Dreiding, Helv.Chim.Acta **55**, 702 (1972).
[34] D. J. Haywood, S.T. Reid, J.Chem.Soc., Perkin Trans. I **1977**, 2457.
[35] G. Stork, T. L. Macdonald, J.Am.Chem.Soc. **97**, 1264 (1975).
[36] E. Yoshii, S. Kimoto, Chem.Pharm.Bull.Jpn. **17**, 629 (1969).
[37] D. N. Reinhoudt, C. G. Leliveld, Tetrahedron Lett. **1972**, 3119.
[38] D. N. Reinhoudt, C. G. Kouwenhoven, Rec.Trav.Chim.Pays-Bas **92**, 865 (1973).
[39] H. Wamhoff, G. Haffmanns, H. Schmidt, Chem.Ber. **116**, 1691 (1983).
[40] H. Wamhoff, G. Haffmanns, Chem.Ber. **117**, 585 (1984).
[41] H. Wamhoff, J. Hartlapp, Chem.Ber. **109**, 1269 (1976). 48
[42] H. Ardill, R. Grigg, V. Sridharan, J. Malone, J.Chem.Soc., Chem.Commun. **1987**, 1296.
[43] D.W. Boerth, F. A. Van-Catledge, J.Org.Chem. **40**, 3319 (1975).
[44] L. Andersen, C.-J. Aurell, B. Lamm, R. Isaksson, J. Sandström, K. Stenvall, J.Chem.Soc., Chem.Commun. **1984**, 411.

[45] G.W. Visser, W. Verboom, D. N. Reinhoudt, S. Harkema, G. J. van Hummel, J.Am.Chem.Soc. **104**, 6842 (1982).
[46] N. G. Rondan, K. N. Houk, J.Am.Chem.Soc. **107**, 2099 (1985).
[47] S. L. Schreiber, C. Santini, Tetrahedron Lett. **22**, 4651 (1981).
[48] T. Shimo, K. Somekawa, J. Kuwakino, H. Uemura, S. Kumamota, O. Tsuge, S. Kanemasa, Chem.Lett. **1984**, 1503.
[49] J. Ficini, A. Duréault, Tetrahedron Lett. **1977**, 809.
[50] P. F. King, L. A. Paquette, Synthesis **1977**, 279.
[51] T. Sano, J. Toda, Y. Horiguchi, K. Imafuku, Y. Tsuda, Heterocycles **16**, 1463 (1981).
[52] G. H. Lange, M.-A. Huggins, E. Neidert, Tetrahedron Lett. **1976**, 4409.
[53] M. Karpf, A. S. Dreiding, Helv.Chim.Acta **58**, 2409 (1975).
[54] S. Hünig, H. Hoch, Chem.Ber. **105**, 2197 (1972).
[55] W. Mock, M. Hartman, J.Am.Chem.Soc. **92**, 5767 (1970).
[56] P. Caubere, G. Guillaumet, M. S. Mourad, Tetrahedron **29**, 1857 (1973).
[57] P. Caubère, Topics Curr.Chem. **73**, 50 (1978).
[58] M. P. Cava, K. Muth, J.Am.Chem.Soc. **82**, 652 (1960).
[59] P. Caubère, G. Guillaumet, M. S. Mourad, Tetrahedron Lett. **1971**, 4673.
[60] P. Caubère, M. S. Mourad, G. Guillaumet, Tetrahedron **29**, 1851 (1973).
[61] M.-C. Carré, P. Caubère, G. Trockle, M. Jacque, Eur.J.Med.Chem., Chim.Therapeut. **12**, 577 (1977).
[62] P. Caubere, L. Lalloz, J.Org.Chem. **40**, 2853 (1975).
[63] M.-C. Carre, B. Gregoire, P. Caubere, J.Org.Chem. **49**, 2050 (1984).
[64] M.-C. Carre, B. Jamart-Gregoire, P. Geoffroy, P. Caubere, Tetrahedron **44**, 127 (1988).
[65] M. S. Manhas, S. G. Amin, A. K. Bose, Heterocycles **5**, 669 (1976).

V. The Cope Rearrangement, the [1.3] Sigmatropic Shift, the Sommelet-Hauser Reaction, and Sulfur-Mediated Ring Expansions

V.1. The Cope Rearrangement

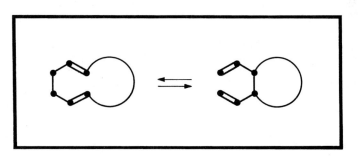

If a compound, containing an 1,5-diene system is heated, it can isomerize in a [3.3] sigmatropic rearrangement [1]. Isomerizations of open chain systems are more familiar than those of alicyclic rings. A ring enlargement can be expected to occur if an alicyclic 1,5-diene, substituted as shown on the top of this page, is heated. In case of a [3.3] shift the reaction proceeds with an enlargement of the ring by four members. Some simple examples of alicyclic systems are collected in Scheme V/1 [2] [3]. In these examples the starting materials as well as the products are 1,5-dienes. This means that both types of compounds can undergo a [3.3] sigmatropic rearrangement.

Depending on the ring size of each the equilibrium can be more to the left or to the right. Some of the carbon atoms in the examples shown in Scheme V/1 can be replaced by oxygen or nitrogen atoms, thus producing heterocyclic systems. A number of factors influence the ratio of alternative products, *e.g.* effects of (methyl) substitution at the terminal carbon atoms of the vinyl groups [9] [10] [11]. The transition state of the uncatalyzed Cope rearrangement is six-centered and can be either boat- or chair-shaped. If both parts of the allylic Cope systems are unsubstituted, the chair geometry is preferred [3]. Not all Cope rearrangements proceed by a cyclic six-centered mechanism; a diradical two-step mechanism may be preferred in some substrates [12].

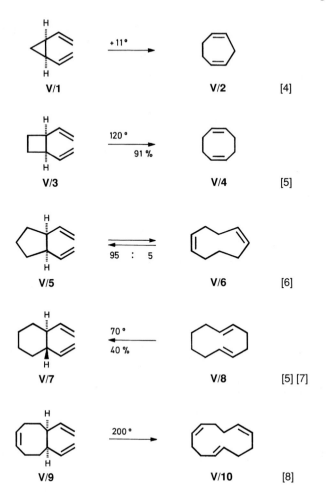

Scheme V/1. The Cope rearrangement of 1,2-divinylcycloalkanes.

The major disadvantages in the application of the Cope rearrangement to the synthesis of organic molecules is the equilibrium between starting material and ring enlargement product. The ratio of the two products is not predictable, *a priori*.

In these cases a modification of the rearrangement, the so-called oxy-Cope rearrangement, is preferred. Thermolysis of a 3-hydroxy-1,5-diene results in the expected 1,5-diene system, but one of the olefinic bonds formed in this process is an enol, which can tautomerize to the corresponding ketone. Thus, a reverse Cope rearrangement cannot take place. Examples are shown in Scheme V/2 [13]. This reaction sequence has been investigated using the thermal behavior of two 1,2-divinylcyclohexanols as model compounds. The *trans*-iso-

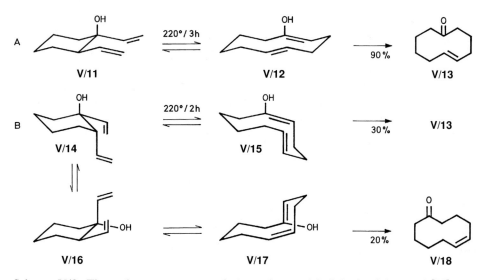

Scheme V/2. Thermal rearrangement of *cis*- and *trans*-1,2-divinylcyclohexanol [13].

mer **V/11**, when heated 3 h at 220°, gave only one product in 90 % yield, (*E*)-5-cyclodecenone (**V/13**). The *cis*-isomer **V/14**, heated under the same conditions for only 2 h, yielded a mixture containing 60 % of **V/13** and 40 % of (*Z*)-5-cyclo-decenone (**V/18**). The total yield of both was approx. 50 %. The formation of

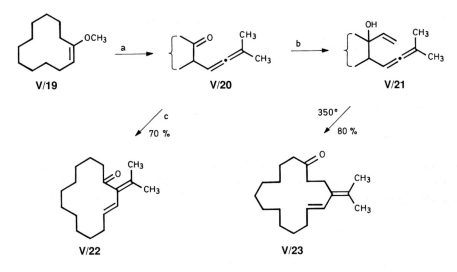

Scheme V/3. Transformation of a 12- to a 16-membered ring by an oxy-Cope rearrangement [16].

a) HC≡C-C(CH₃)₂OH, p-TsOH, benzene, Δ
b) CH₂=CHMgBr, THF   c) (C₂H₅)₂O, hν.

two products in the latter reaction has been explained as taking place through two different chair conformations, **V/15** and **V/17**, in the transition state of **V/14** [13], compare [14] [15]. A variation of the oxy-Cope rearrangement with the vinyl allenyl alcohol, **V/21**, prepared from cyclododecanone by the method outlined in Scheme V/3, has been also described [16]. The alcohol, **V/21**, underwent a thermal Cope rearrangement which resulted in the 16-membered ketone, **V/23**, in high yield. A prolonged irradiation of the synthetic precursor of **V/21**, the allenic ketone **V/20**, produced an (n + 2) enlargement by an 1,3-acyl migration.

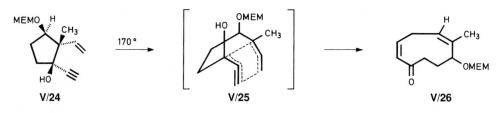

Scheme V/4.

One of the double bonds in the Cope system can be replaced by a triple bond, Scheme V/4. The five-membered **V/24**, heated to 170° for 20 min only, gave the nine-membered **V/26**, containing a (Z)-double bond and an α,β- unsaturated carbonyl group in a yield of 62 % [17]. This reaction step was part of a synthesis of phoracantholide I (compare Chapter VII.3).

If the oxy-Cope rearrangement is carried out with divinyl diols of type **V/27** or **V/30** (Scheme V/5), ring-enlarged 1,6-diketones of type **V/28** or **V/31** are produced [18] [19] [20] [21]. Depending on the ring size of the resulting dike-

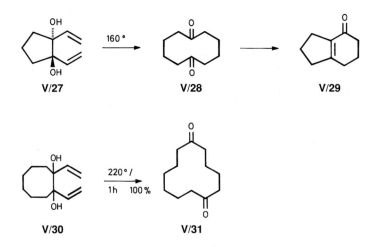

Scheme V/5. Di-oxy-Cope rearrangement [19] [20].

tones and depending on the reaction conditions and workup procedures used for their preparation, additional products are observed. For example: *trans*-1,2-cyclopentanediol forms the aldol condensation product, **V/29**, by loss of water from the first formed Cope product, **V/28** [20]. The Cope rearrangement product will be stable if the newly formed ring is a large ring (*e.g.* **V/31**) and not a medium sized one. The latter undergoes easily transannular reactions [19].

The oxy-Cope rearrangement is frequently used as a key step, or at least as an important reaction, in the synthesis of natural and unnatural compounds containing medium sized rings. The sesquiterpenes acoragermacrone (**V/37**) and preisocalamendiol (**V/35**), each in their racemic forms, have been synthesized by application of the oxy-Cope reaction, (Scheme V/6) [22]. The monocyclic six-membered monoterpene isopiperitenone (**V/32**) was the starting material. Treatment of **V/32** with 2-lithio-3-methyl-1-butene gave the alcohol **V/33**, which was immediately subjected to conditions [23] of the oxy-Cope rearrangement. The ten-membered ring compound, **V/34**, was formed in 73 % yield. In order to isomerize the (*Z*)-double bond in **V/31** to its (*E*)-isomer in **V/37**, a kinetic 1,4-addition/elimination sequence was developed. The experimental details are given in Scheme V/6. Alternatively, **V/34** could be deconjugated by kinetic enolate protonation to yield preisocalamendiol (**V/35**) [22][1)].

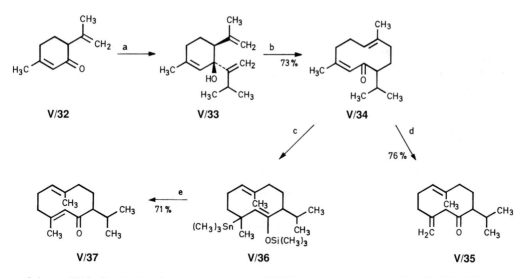

Scheme V/6. Synthesis of acoragermacrone (**V/37**) and preisocalamendiol (**V/35**) [22].

a) $LiC(=CH_2)$-$CH(CH_3)_2$, THF, $-78°$ b) KH, THF, [18]crown-6, $25°$
c) $(CH_3)_3SnLi$, $(CH_3)_3SiCl$
d) lithiumdiisopropylamide, THF, hexamethylphosphoramide, HOAc
e) MnO_2, CH_2Cl_2, 30 min.

---

1) For an alternative approach to the germacranes by an oxy-Cope rearrangement see ref. [25].

The oxy-Cope rearrangement has also been used in synthesis of (±)-peripla-none-B (**V/38**), the sex excitant pheromone of the American cockroach (*Periplaneta americana*) [24]. Scheme V/7 gives the preparation of the starting material, **V/43**, for the rearrangement step. The divinylcyclohexenol derivative, **V/43**, smoothly underwent an oxy-Cope rearrangement after conversion to its potassium salt. The reaction mixture was cooled to −78°, treated with $(CH_3)_3SiCl$ and finally oxidized with *m*-chloroperbenzoic acid [24].

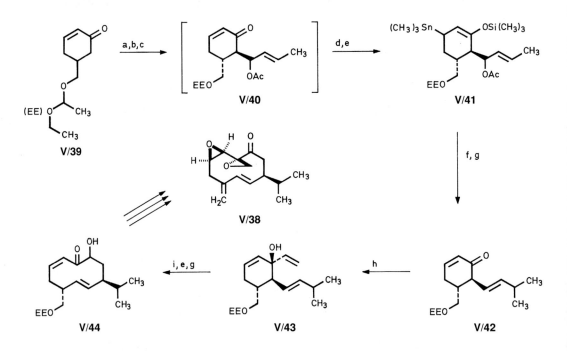

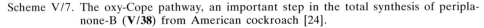

Scheme V/7. The oxy-Cope pathway, an important step in the total synthesis of peripla-none-B (**V/38**) from American cockroach [24].

a) Lithiumdiisopropylamide, THF, 0°   b) $CH_3CH=CH-CHO$, −78°
c) Ac_2O, −78°   d) $(CH_3)_3SnLi$, −78°   e) $(CH_3)_3SiCl$, −78°
f) $Li(CH_3)_2Cu$, $(C_2H_5)_2O$, 0°   g) *m*-chloroperbenzoic acid
h) $LiCH=CH_2$, $(C_2H_5)_2O$ i) KH, THF, [18]crown-6, 1 h, 70°.

Attempts have been made to find reaction sequences which allow the introduction of more than four atoms into a ring by a Cope rearrangement. Two of these methods should be mentioned, both quite different. The first method uses an "enlarged Cope system", which forms bigger rings than the normal Cope system. In the second method the product of one Cope rearrangement can be easily transformed into the starting material for a second Cope shift sequence.

The principal example of the macroexpansion methodology is given in Scheme V/8. Treatment of 1,2-(E,E)-di(buta-1,3-dienyl)cyclohexanol (**V/47**) with potassium hydride in tetrahydrofuran resulted in the formation of the 14-membered ring enolate, which provided cyclotetradeca-3,5,7-trien-1-one, **V/50**, when protonated. Whether the reaction involves two consecutive [3.3] or one [5.5] sigmatropic rearrangements will not be discussed here [26] [27] [28]. Eight atoms can be incorporated in one step, a reaction path which belongs indeed to the largest known "macroexpansion". The synthesis of the starting material, **V/47**, is outlined in Scheme V/8. This expansion methodology was also used for the conversion of 7-chloro-3-methyl-3-cycloheptenone into the 15-membered muscone [29].

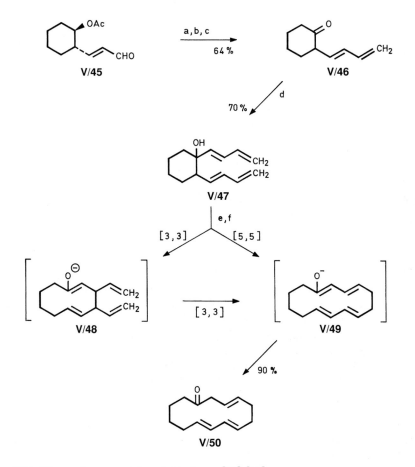

Scheme V/8. Ring enlargement by eight atoms [26] [27].

a) $(C_6H_5)_3PCH_2$ b) $LiAlH_4$ c) CrO_3 d) $LiCH=CH-CH=CH_2$
e) KH, THF, 20°, 1 h   f) NH_4Cl.

Scheme V/9 shows the essentials of a second approach, the so-called "repeatable ring expansion reaction", and its application to the cyclododecanone series [2]. The α,β-unsaturated β-ketoester, **V/52**, was prepared from cyclododecanone. Its α,β-divinylation was carried out in two steps. A Cope rearrangement under the reaction conditions of α-vinylation[2) led to the formation of the 16-membered **V/55**. The reaction product **V/55** can be transformed using similar procedures, to analogues of the starting materials **V/51** or **V/52** [2]. It is possible to enlarge these compounds by similar methods to even larger rings. Thus, one has a repeatable sequence for ring enlargement.

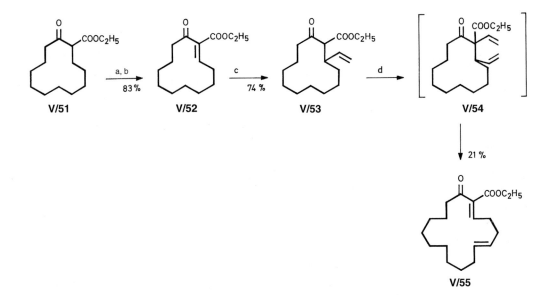

Scheme V/9. A repeatable version of the Cope rearrangement [2]. The configuration of the double bonds in **V/55** is unknown.

a) NaH, C_6H_5SeCl b) H_2O_2 c) $CH_2=CHMgBr$, CuI
d) $C_6H_5\text{-}S(O)CH=CH_2$, NaH, THF, heat.

A method of "repetitive ring expansion" of cyclic ketones was published based on the use of (phenylseleno)acetaldehyde on the siloxy-Cope rearrangement [29a]. The authors were able to transform cyclododecanone into cycloeicosadec-5-en-1-one in 23 % yield.

---

2) (Phenylseleno)acetaldehyde was recommended as an alternative synthetic equivalent to the vinyl carbocation for α-vinylation of ketones [30].

V.2. [1.3] Sigmatropic Shift – A Method of Ring Enlargement

A ring expansion by two carbon atoms was discovered on heating (Z)-1-vinyl-cyclonon-3-en-1-yl-trimethylsilyl ether (**V/62**), Scheme V/10 [31] [32]. Two types of trimethylsiloxy enol ethers were observed, each formed by a [1.3] or a [3.3] sigmatropic shift which, after hydrolysis, gave the ketones **V/63**, **V/64**, and **V/65**. The protection of the tertiary alcohol function is not necessary, if the reaction is done in the presence of potassium hydride [33] [34]. The reaction proceeds when the cyclic 1-vinyl alcohols have either a double bond or a benzo group at the 3-position. Other examples of this reaction type are given in Scheme V/10.

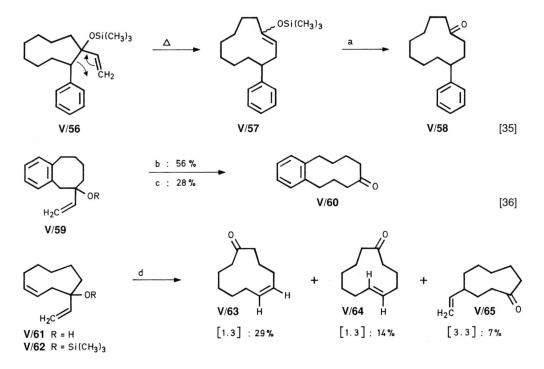

Scheme V/10. [1.3] Sigmatropic shift as a tool of ring expansion by two atoms.

a) Hydrolysis b) R=K; 20°, 5.5 h c) R=Si(CH₃)₃, 350°, hydrolysis
d) R=H, KH, hexamethylphosphorous triamide, 25°, 5.5 h.

This ring expansion has been applied to the synthesis of the 15-membered (±)-muscone [37]. First cyclododecanone was tranformed to cyclotridec-3-enone (**V/67**) in a five step synthesis [38]. The latter, treated with the Grignard reagent formed from prop-1-enyl bromide, generated a mixture of the isomeric compounds, **V/68**, in nearly 50 % yield. Obviously the methyl group sterically

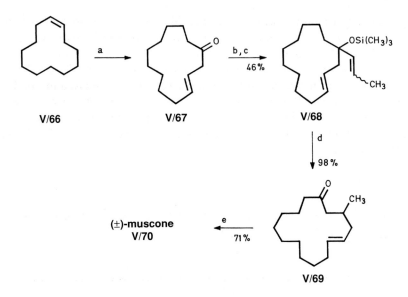

Scheme V/11. Synthesis of (±)-muscone (**V/70**) from cyclododecene (**V/66**) including a [1.3]
sigmatropic shift ring expansion [37].

a) Compare ref. [38]   b) H₃CCH=CHMgBr   c) (CH₃)₃SiCl   d) 320°, 5 h
e) H₂/Pt.

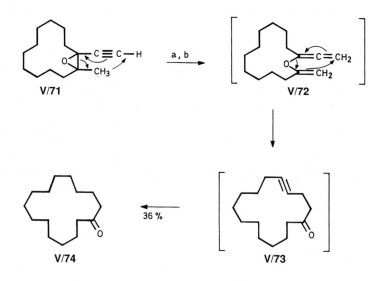

Scheme V/12. Formation of cyclopentadecanone (**V/74**) from cyclododecanone [39].
a) 600°/0.25 torr   b) H₂, 10, Pd/C, C₂H₅OH.

destabilizes the transition state for the [3.3] shift more than that for the [1.3] shift. The resulting 15-membered compound **V/69** was hydrogenated to (±)-muscone (**V/70**) [37], Scheme V/11.

A cyclopentadecanone (**V/74**) and a (±)-muscone (**V/70**) synthesis were carried out by another method related to the reactions discussed in this Chapter. Scheme V/12 shows the thermal transformation of the cyclododecanone derivative, **V/71**, to **V/74** [39]. The proposed mechanism for this conversion presumably includes a [1.5] hydrogen shift, followed by a [3.3] sigmatropic rearrangement.

V.3. Sommelet-Hauser Rearrangement and Sulfur-Mediated Ring Expansion

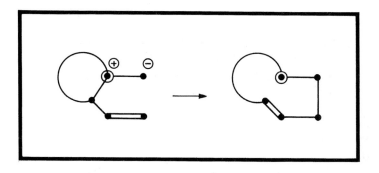

The Sommelet-Hauser rearrangement was originally named as an "ortho substitution rearrangement" [40]. If 1,1-dimethyl-2-phenylpiperidinium iodide (**V/75**) was treated with sodium amide in liquid ammonia, the nine-membered benzannelated (an ortho fused ring) tertiary amine, **V/78**, was obtained in a yield of 83 %. The proposed mechanism, including the formation of the ylide, **V/76**, is shown in Scheme V/13. Similar reactions have been carried out with four- and five-membered α-phenyl ammonium compounds. It seems that using phenyllithium as a base gives lower yields of the ring enlargement products than amide/liquid ammonia, Scheme V/14. There exist also examples involving rearrangement of an α-vinyl instead of an α-phenyl ammonium salt [41] [42].

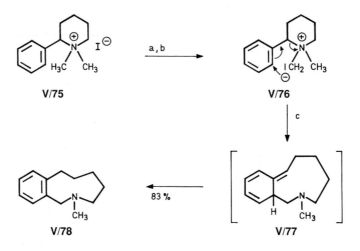

Scheme V/13. The Sommelet-Hauser rearrangement of 1,1-dimethyl-2-phenylpiperidinium
iodide (**V/75**) [40].

a) NaNH$_2$, NH$_3$ liq.   b) NH$_4$Cl-H$_2$O   c) [2.3] shift.

These rearrangement reactions are interpretable in terms of [2.3] sigmatropic
shifts of the intermediate ylides. A number of such rearrangements of open-
chain systems have been described, involving sulfonium ylides [43] [44] [45],
ammonium ylides [46] [57], anions in α-position to oxygen (Wittig rearrangement) [48] [49], and fluorenyl carbanions [50].

An analogous reaction is observed, when the β-lactam, **V/88**, reacts with
lithium diisopropylamide in tetrahydrofurane, Scheme V/15. The seven-membered lactam **V/89** was obtained in a virtually quantitative yield. The driving
force of this reaction depends on the release of ring strain in changing from a
four- to a seven-membered ring, and on the formation of a resonance stabilized
secondary amide. The mono-methylated analogue **V/90**, treated with the same
reagents, gave the five-membered **V/92**, as well as the expected seven-membered **V/91**. This reaction sequence can be explained by the homolytic cleavage
of the benzylic carbanion to the diradical anion intermediate **V/93**, which
recombines, to give, mainly, the five-membered ring product **V/92**. At low
temperature, only the benzylic anion is created from the tetramethyl-derivative,
V/94, which gave **V/95**, on addition of D$_2$O to the basic solution. At higher
temperature two products (90% yield, ratio 1:1, 3,3,4,4-tetramethyl-5-phenyl-2-
pyrrolidone and 2,2,3-trimethyl-3-butenamide) are observed whose structures
can be explained by homolysis of the activated C,N bonds in **V/94**, Scheme
V/15 [42].

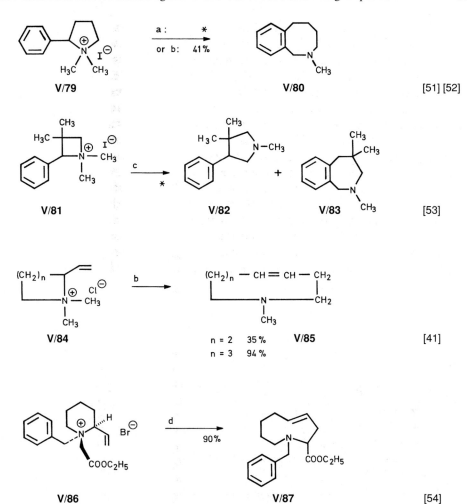

Scheme V/14. Further examples of the Sommelet-Hauser rearrangement.

a) C_6H_5Li, $(C_2H_5)_2O$, 20°, 5 d   b) $NaNH_2$, NH_3 liq.

c) C_6H_5Li, $(C_2H_5)_2O$, 10°   d) 1,8-diazabicyclo[5.4.0]undec-7-ene, THF.

* Yields of both compounds are very low.

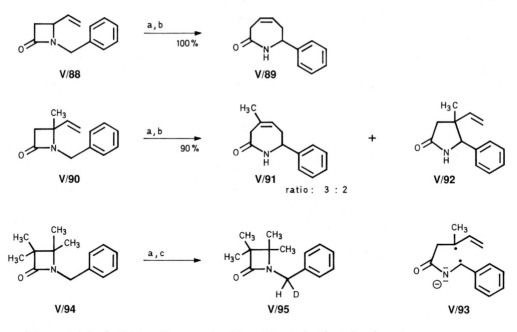

Scheme V/15. a) Lithiumdiisopropylamide, THF, –78°   b) H$_2$O   c) D$_2$O.

The examples given in Scheme V/15 demonstrate the strong dependence on structure variations, temperature and the nature of the cation.

The Sommelet-Hauser rearrangement has rarely been used as a tool in organic synthesis and then only for very special systems [54]. The reason is undoubtedly that the Hofmann elimination and the dealkylation reaction of quaternary nitrogen atoms are in direct competition with the ring expansion reactions of these substrates.

Sulfur ylides can undergo a [2.3] sigmatropic rearrangement which has been very useful in organic synthesis. Scheme V/16 shows some fundamental applications of this reaction.

The synthesis of the key compound 2-vinyltetrahydrothiophene (**V/97**) was accomplished by base catalyzed cyclisation of **V/96** or Grignard vinylation of the 2-chloro-thiacycloalkanes [51]. In the next step, the sulfide was alkylated to get the sulfonium salt, which was then deprotonated to obtain the corresponding sulfur ylide. For alkylation, the copper-catalyzed diazo decomposition path (shown in the conversion **V/97** → **V/100**) was not very efficient [55] [56]. Alkylation with allylbromides [57] or, better with triflates [58] gave good results. It is most important that the derived sulfonium salts are not dealkylated by the non-nucleophilic triflate ion. Triflates are easily prepared from the corresponding alcohols and triflic anhydride [59]. Stabilized sulfur ylides have been prepared to avoid strongly basic conditions. The eight-membered ring compounds, **V/99**,

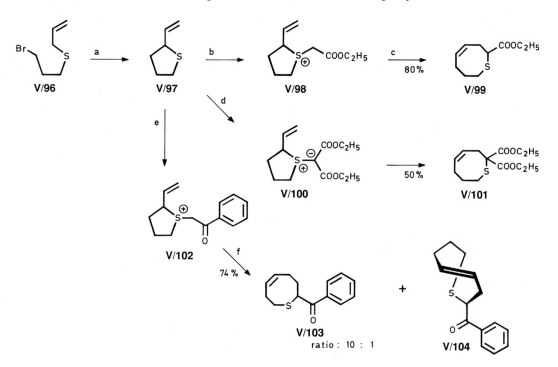

Scheme V/16. Ring enlargement of sulfur ylides [55].

    a) Lithiumdiisopropylamide b) $CF_3SO_3CH_2COOC_2H_5$
    c) KOtBu d) $N_2C(COOC_2H_5)_2$, Cu e) $CF_3SO_3CH_2C(=O)C_6H_5$
    f) 1,8-diazabicyclo[5.4.0]undec-7-ene.

V/101, and **V/103,** have been prepared in 80, 50, and 74 % yield, respectively, by [2.3] sigmatropic rearrangement of the ylides, see Scheme V/16. The ring expansion of a mixture of *trans-* and *cis*-1-ethyl-2-vinylthiolanium hexafluorophosphates (**V/105**) gave a mixture of three sulfides, (*Z*)-2-methyl-thiacyclooct-4-ene (**V/106**) and the two diastereoisomerically related (*E*)-(SR,RS)- and (*E*)-(RR,SS)-2-methylthiacyclooct-4-enes (**V/107**) and (**V/108**)[3], Scheme V/17. The existence of the two (*E*) isomers is evidence for the structure of the molecule, holding two elements of chirality, a chiral center, and a plane. The diastereoisomers are stabile because of restricted conformational inversion around the chiral plane [45] [60].

    Through sulfur ylide reactions rings can be enlarged by three atoms, two from the vinyl side chain and a third one from the α-alkyl substituent of the sulfonium ion. The reaction product is, like the starting material, a thiacycloalkane

---

3) Both **V/107** and **V/108** undergo irreversible thermal conversion to **V/106** ($\approx$ 100°, several hours).

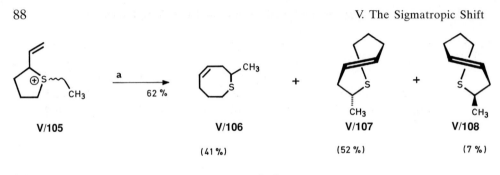

V/105   **V/106**   **V/107**   **V/108**

                          (41 %)   (52 %)   (7 %)

Scheme V/17. Sulfonium ylide rearrangement [60].

a) KOtBu, THF, –40°.

and, it is important to note, is again α-substituted. These considerations allow the use of the product of the first ring enlargement as the starting material for a second enlargement step. This idea is shown in Schemes V/18 and V/19. The treatment of the sulfonium salt **V/111** carrying an appropriate S-allyl substituent gave the nine-membered thiacycloalkene **V/112** in the presence of potassium *tert*-butoxide. Compound **V/112** has again a vinyl substituent in an α-position to the ring sulfur atom. The twelve-membered **V/113** was already known [61]. Using this ring growing technique [55] [62], the "repeatable ring expansion method" has been developed. It can be used to synthesize the 17-membered compounds, **V/118** and **V/119**, from five-membered **V/97**, Scheme V/19 [57] [57a]. Alkylation with allyl bromide in 2,2,2-trifluoroethanol followed by potas-

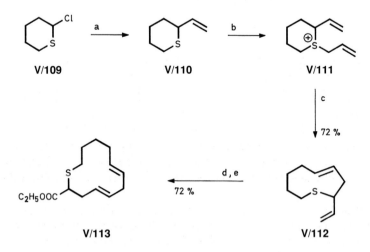

V/109   **V/110**   **V/111**

                                                       c

                                                       72 %

V/113   **V/112**

Scheme V/18. Ring transformations from six- via nine- to twelve-membered ring using the "ring-growing" technique [61].

a) $CH_2=CH-MgX$ b) $CF_3SO_3CH_2-CH=CH_2$ c) KOtBu
d) $CF_3SO_3CH_2-COOC_2H_5$ e) 1,8-diazabicyclo[5.4.0]undec–7-ene.

sium hydroxide treatment under phase transfer conditions gave the ring enlargement products. In four reaction steps (4 x 3 carbon atoms = 12 additional carbon atoms as ring members), the five-membered starting material, **V/97**, has been converted to the two 17-membered isomeric compounds **V/118** and **V/119**. The (*E,Z*)-geometry of the compounds, given in Scheme V/19, agrees with the literature [57].

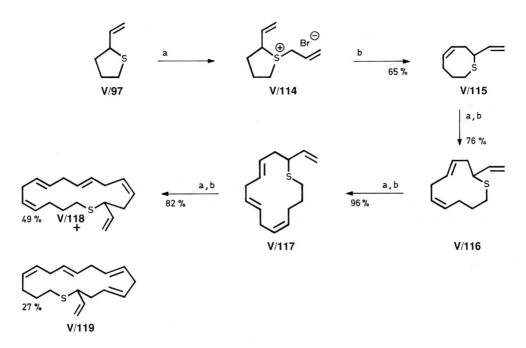

Scheme V/19. Ring transformations from five- via eigth-, eleven- and 14- to 17-membered ring using the "repeatable ring expansion method" [57] [57a].

a) H₂C=CH-CH₂Br, CF₃CH₂OH   b) KOH, H₂O, pentane.

The interest in medium and large ring thiacarbocycles is not sufficient to justify intensive synthetic research, but, since there are specific methods to eliminate sulfur, cyclic sulfides might be converted into other important ring systems. Thus the Ramberg-Bäcklund sulfur extrusion method, can be used to convert a thiacycloalkane to the corresponding cycloalkene. This requires the transformation of the sulfide to an α-halo sulfone. The base catalyzed mechanism involves the formation of an episulfone and elimination of sulfur dioxide and olefin. An example of this reaction, the conversion of **V/120** to **V/123**, is given in Scheme V/20. It should be noted that the Ramberg-Bäcklund reaction leads to a ring contraction by one member. The formation of the cycloalkene is observed in high yield [63].

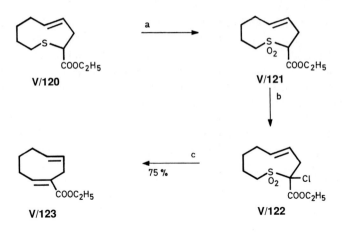

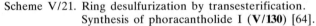

Scheme V/20. Ramberg-Bäcklund sulfur extrusion [57a] [63].

a) *m*-Chloroperoxybenzoic acid   b) NaH, dimethoxyethane/C₆Cl₆   c) heat.

Another way to desulfur the ring itself is to initiate an additional ring enlargement triggered by a substituent of the ring. An illustrative example, the synthesis of phoracantholide I (**V/130**), is given in Scheme V/21, see Chapter VII.3.

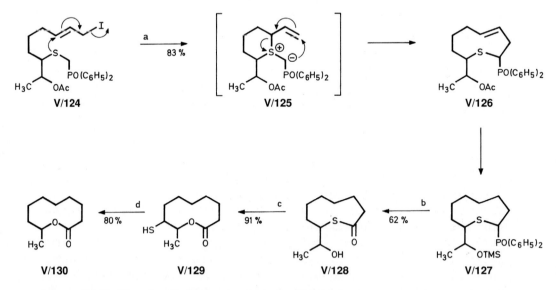

Scheme V/21. Ring desulfurization by transesterification.
          Synthesis of phoracantholide I (**V/130**) [64].

a) K₂CO₃, CH₃CN, heat   b) BuLi, O₂, H₂O.
c) camphorsulfonic acid, CH₂Cl₂
d) Bu₃SnH, azabisisobutyronitrile (AIBN).

The starting material, **V/124**, was heated with potassium carbonate in methylene chloride. The reaction proceeds to **V/126** presumably *via* the six-membered sulfonium salt **V/125**, followed by a [2.3] sigmatropic shift. The double-bond in **V/125** was reduced (diimide), and the protecting group was converted to give **V/127**, which was transformed into the thiolactone, **V/128**, by a variation of Horner-Bestmann oxygenation. The translactonization was carried out in the presence of camphorsulfonic acid to yield the lactone, **V/129**. Desulfurization was achieved by heating **V/129** with 2.1 equiv. of tributyl-tinhydride/AIBN[4]) [64]. Further results of the translactonization are discussed in refs. [58] [64]. The two examples, given in Scheme V/22, show that the thermodynamic advantage of O-acyl relative to S-acyl is comparable to the strain in the eight-membered ring relative to a six- or seven-membered ring.

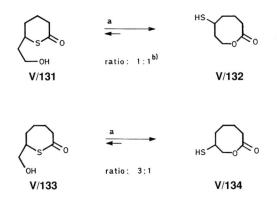

V/131   ratio: 1:1[b)]   V/132

V/133   ratio: 3:1   V/134

Scheme V/22. Thiolactone – lactone rearrangement [58].

a) Camphorsulfonic acid   b) The reaction of **V/131**, **V/132** is not a true equilibrium because of decomposition. Yield of the products: first reaction 20 %, second reaction 70 %.

The sulfide bridge in medium sized rings can be used in stereocontrolled syntheses of complex molecules [65], (Chapter IX.1), for example the synthesis of zygosporin E [66] and the synthesis of cytochalasans [67].

S-Imides of 2-vinylthiacycloalkanes (Scheme V/23) undergo [2.3] sigmatropic shifts to form azathia-ring expansion products. The eight- and nine-membered compounds **V/138** and **V/139** which are formed at higher temperatures (xylene, reflux) have a (Z)-geometry of the double bond. The ten-membered **V/140**, with a (E)-double bond, is generated directly during chloramine T treatment of 2-vinylthiacycloheptane [68].

---

4) This procedure has been found to reduce other secondary mercaptans in reasonable yield [64].

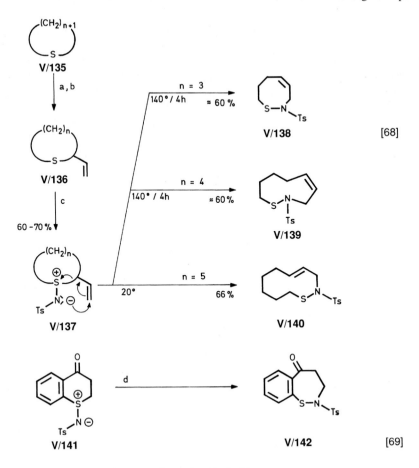

Scheme V/23. Ring expansion reaction of sulfurimides.

a) N-Chlorosuccinimid   b) CH₂=CH-MgBr   c) Ts-NClNa, CH₃OH, 20°
d) N(C₂H₅)₃, CHCl₃.

A reaction leading to similar heterocyclic systems was presented earlier: The
thiachroman-4-on-S-imide, **V/141**[5], gave benzothiazepin-5-one, **V/142**, when
treated with base. Presumably a Hofmann elimination followed by cyclization
is responsible for the formation of **V/142**. Two further examples are listed in
Scheme V/24.

Ketene, dichloroketene, and related compounds react with allylic ethers, sulfides, and selenides in a [3.3] sigmatropic rearrangement. This interesting reaction allows the synthesis of a number of medium and large ring compounds by
expansion [72] [73] [74]. Dichloroketene has been prepared *in situ* by slow addi-

_____
5) 1-(p-Tolylsulphonylimino)-1-thiachroman-4-one.

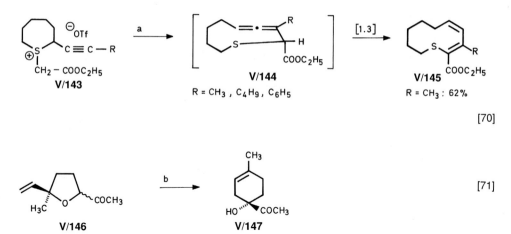

Scheme V/24. Further examples of [2.3] sigmatropic shift ring enlargements.

a) 1,8-Diazabicyclo[5.4.0]undec-7-ene, CH₃CN, 20°, 15 min   b) KOtBu.

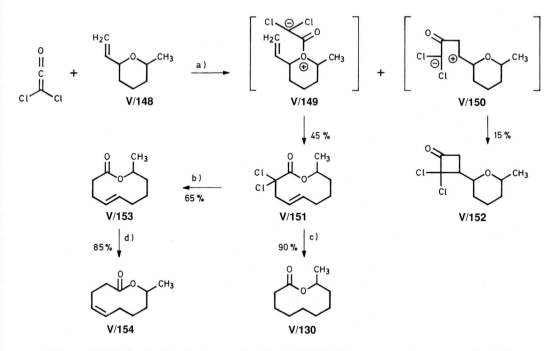

Scheme V/25. Synthesis of phoracantholide I (**V/130**) by a type rearrangement involving
1,3-dipolar intermediates [72] [74].

a) $(C_2H_5)_2O$, 25°, 4 h   b) Zn, HOAc, 80°, 6 h
c) $H_2/5\%$ Pd-C/dioxane/pyridine   d) $(C_6H_5S)_2$, hexane, hν, 20°.

tion of trichloroacetylchloride to a suspension of copper activated zinc powder in ether solution. Allylic O-, S-, and Se-ethers form 1,3-dipolar intermediates with dichloroketene. The reaction of dichloroketene and the tetrahydropyrane derivative **V/148** is shown in Scheme V/25.

The 1,3-dipolar intermeditate **V/149** is then transformed to the ten-membered dichlorolactone **V/151** by a [3.3] sigmatropic rearrangement. The side product, **V/152**, is obtained by a [2+2] cycloaddition [72] [74]. Under standard conditions, the dichlorolactone, **V/151**, was converted to phoracantholide I (**V/130**). Dechlorination of **V/151** gave **V/153** which was converted to phoracantholide J, **V/154**, by irradiation. Both phoracantholides are occurring naturally, see Chapter VII.2. The same technique has been applied to the synthesis of other products [73] [74].

References

[1] S. J. Rhoads, N. R. Raulins, Org. Reactions **22**, 1 (1975).
[2] J. Bruhn, H. Heimgartner, H. Schmid, Helv.Chim.Acta **62**, 2630 (1979).
[3] W. v. E. Doering, W. R. Roth, Tetrahedron **18**, 67 (1962).
[4] M. P. Schneider, J. Rebell, J.Chem.Soc., Chem.Commun. **1975**, 283.
[5] E. Vogel, Angew.Chem. **74**, 829 (1962).
[6] E. Vogel, W. Grimme, E. Dinné, Angew.Chem. **75**, 1103 (1963).
[7] C. A. Grob, H. Link, P.W. Schiess, Helv.Chim.Acta **46**, 483 (1963).
[8] R. Rienäcker, N. Balcioglu, Liebigs Ann.Chem. **1975**, 650.
[9] J. A. Berson, P. B. Dervan, J.Am.Chem.Soc. **94**, 7597 (1972).
[10] J. A. Berson, P. B. Dervan, J. A. Jenkins, J.Am.Chem.Soc. **94**, 7598 (1972).
[11] G. S. Hammond, C. D. DeBoer, J.Am.Chem.Soc. **86**, 899 (1964).
[12] J. March, Advanced Organic Chemistry, 3.Ed. John Wiley & Sons, New York 1985.
[13] E. M. Marvell, W. Whalley, Tetrahedron Lett. **1970**, 509.
[14] N. Bluthe, M. Malacria, J. Gore, Tetrahedron Lett. **23**, 4263 (1982).
[15] M. Nishino, H. Kondo, A. Miyake, Chem.Lett. **1973**, 667.
[16] R. C. Cookson, P. Singh, J.Chem.Soc. C **1971**, 1477.
[17] T. Ohnuma, N. Hata, N. Miyachi, T. Wakamatsu, Y. Ban, Tetrahedron Lett. **27**, 219 (1986).
[18] F. Brown, P. Leriverend, J.-M. Conia, Tetrahedron Lett. **1966**, 6115.
[19] P. Leriverend, J.-M. Conia, Bull.Soc.Chim.France **1970**, 1040.
[20] P. Leriverend, J.-M. Conia, Bull.Soc.Chim.France **1970**, 1060.
[21] E. N. Marvell, T. Tao, Tetrahedron Lett. **1969**, 1341.
[22] W. C. Still, J.Am.Chem.Soc. **99**, 4186 (1977).
[23] D. A. Evans, A. M. Golob, J.Am.Chem.Soc. **97**, 4765 (1975).
[24] W. C. Still, J.Am.Chem.Soc. **101**, 2493 (1979).
[25] S. L. Schreiber, C. Santini, Tetrahedron Lett. **22**, 4651 (1981).
[26] P. A. Wender, S. M. Sieburth, Tetrahedron Lett. **22**, 2471 (1981).
[27] P. A. Wender, S. M. Sieburth, J. J. Petraitis, S. K. Singh, Tetrahedron **37**, 3967 (1981).
[28] A. G. Angoh, D. L. J. Clive, J.Chem.Soc., Chem.Commun. **1984**, 534.
[29] P. A. Wender, D. A. Holt, S. M. Sieburth, J.Am.Chem.Soc. **105**, 3348 (1983).
[29a] D. L. J. Clive, A. G. Angoh, S. C. Suri, S. N. Rao, C. G. Russell, J.Chem.Soc., Chem. Commun. **1982**, 828.

[30] D.L. J. Clive, C. G. Russell, S. C. Suri, J.Org.Chem. **47**, 1632 (1982).

[31] R.W. Thies, J.Chem.Soc., Chem.Commun. **1971**, 237.

[32] R.W. Thies, J.Am.Chem.Soc. **94**, 7094 (1972).

[33] R.W. Thies, E. P. Seitz, J.Chem.Soc., Chem.Commun. **1976**, 846.

[34] R.W. Thies, E. P. Seitz, J.Org.Chem. **43**, 1050 (1978).

[35] R.W. Thies, Y. B. Choi, J.Org.Chem. **38**, 4067 (1973).

[36] R.W. Thies, M. Meshgini, R. H. Chiarello, E. P. Seitz, J.Org.Chem. **45**, 185 (1980).

[37] R.W. Thies, K. P. Daruwala, J.Chem.Soc., Chem.Commun. **1985**, 1188.

[38] R.W. Thies, R. E. Bolesta, J.Org.Chem. **41**, 1233 (1976).

[39] M. Karpf, A. S. Dreiding, Helv.Chim.Acta **60**, 3045 (1977).

[40] D. Lednicer, C. R. Hauser, J.Am.Chem.Soc. **79**, 4449 (1957).

[41] B. Hasiak, Compt.rend.Acad.Sci. Paris **C282**, 1003 (1976).

[42] T. Durst, R. v. d.Elzen, M. J. LeBelle, J.Am.Chem.Soc. **94**, 9261 (1972).

[43] J. E. Baldwin, R. E. Hackler, D. P. Kelly, J.Am.Chem.Soc. **90**, 4758 (1968).

[44] B. M. Trost, R. LaRochelle, Tetrahedron Lett. **1968**, 3327.

[45] V. Ceré, C. Paolucci, S. Pollicino, E. Sandri, A. Fava, J.Org.Chem. **46**, 3315 (1981).

[46] G.V. Kaiser, C.W. Ashbrook, J. E. Baldwin, J.Am.Chem.Soc. **93**, 2342 (1971).

[47] R.W. Jemison, W. D. Ollis, J.Chem.Soc., Chem.Commun. **1969**, 294.

[48] V. Rautenstrauch, J.Chem.Soc., Chem.Commun. **1970**, 4.

[49] K. Mikami, S. Taya, T. Nakai, Y. Fujita, J.Org.Chem. **46**, 5447 (1981).

[50] J. E. Baldwin, F. J. Urban, J.Chem.Soc., Chem.Commun. **1970**, 165.

[51] H. Daniel, F. Weygand, Liebigs Ann.Chem. **671**, 111 (1964).

[52] G. C. Jones, C. R. Hauser, J.Org.Chem. **27**, 3572 (1962).

[53] A. G. Anderson, M.T. Wills, J.Org.Chem. **33**, 536 (1968).

[54] E. Vedejs, M. J. Arco, D.W. Powell, J. M. Renga, S. P. Singer, J.Org.Chem. **43**, 4831 (1978).

[55] E. Vedejs, J. P. Hagen, J.Am.Chem.Soc. **97**, 6878 (1975).

[56] E. Vedejs, M. J. Mullins, J.Org.Chem. **44**, 2947 (1979).

[57] R. Schmid, H. Schmid, Helv.Chim.Acta **60**, 1361 (1977).

[57a] R. Schmid, Thesis, University of Zürich, 1978.

[58] E. Vedejs, Acc.Chem.Res. **17**, 358 (1984).

[59] E. Vedejs, D. A. Engler, M. J. Mullins, J.Org.Chem. **42**, 3109 (1977).

[60] V. Ceré, S. Pollicino, E. Sandri, A. Fava, J.Am.Chem.Soc. **100**, 1516 (1978).

[61] E. Vedejs, M. J. Mullins, J. M. Renga, S. P. Singer, Tetrahedron Lett. **1978**, 519.

[62] E. Vedejs, M. J. Arnost, J. P. Hagen, J.Org.Chem. **44**, 3230 (1979).

[63] E. Vedejs, S. P. Singer, J.Org.Chem. **43**, 4884 (1978).

[64] E. Vedejs, D.W. Powell, J.Am.Chem.Soc. **104**, 2046 (1982).

[65] M. Braun, Nachr.Chem.Tech.Lab. **33**, 1066 (1985).

[66] E. Vedejs, J. D. Rodgers, S. J. Wittenberger, J.Am.Chem.Soc. **110**, 4822 (1988).

[67] E. Vedejs, J. G. Reid, J.Am.Chem.Soc. **106**, 4617 (1984).

[68] H. Sashida, T. Tsuchiya, Heterocycles **22**, 1303 (1984).

[69] Y. Tamura, Y. Takebe, S. M. M. Bayomi, C. Mukai, M. Ikeda, M. Murase, M. Kise, J.Chem.Soc., Perkin Trans. I **1981**, 1037.

[70] H. Sashida, T. Tsuchiya, Heterocycles **19**, 2147 (1982).

[71] A. F. Thomas, R. Dubini, Helv.Chim.Acta **57**, 2084 (1974).

[72] R. Malherbe, D. Belluš, Helv.Chim.Acta **61**, 3096 (1978).

[73] E. Vedejs, R. A. Buchanan, J.Org.Chem. **49**, 1840 (1984).

[74] R. Malherbe, G. Rist, D. Belluš, J.Org.Chem. **48**, 860 (1983).

VI. Transamidation Reactions

VI.1. Transamidation Reactions

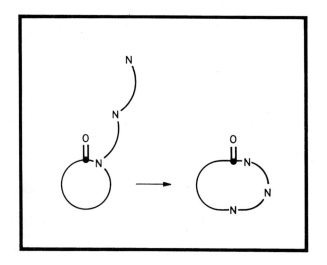

In 1968, the isolation of two new spermidine alkaloids, oncinotine (**VI/1**) and iso-oncinotine (**VI/3**), from the African plant *Oncinotis nitida* Benth. (Apocynaceae) was published Scheme VI/1 [1].

Although a structure was published for **VI/1**, it was not crystallized, despite many attempts at purification. By TLC, samples of oncinotine always seemed to contain some iso-oncinotine (**VI/3**). Subsequently, it was learned that the crude oncinotine contained another isomer, **VI/2**, neo-oncinotine, but no iso-oncinotine. During the purification attempts, it appeared that **VI/2** was converted to **VI/3** (Scheme VI/1) [2]. This reaction has been the starting point of a series of investigations in the field of transamidation reactions.

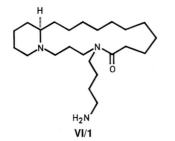

VI/1

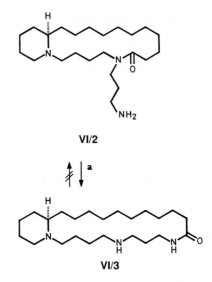

VI/2

a

VI/3

Scheme VI/1. The spermidine alkaloids oncinotine (**VI/1**), neo-oncinotine (**VI/2**), and iso-oncinotine **VI/3**).
a) Base or acid or Δ

Aminoamides such as **VI/4** are instable; they rearrange under acid or base (shown) catalysis to **VI/9**[1]. The mechanism of this reaction is shown in Scheme VI/2 [3]. The driving force of this transamidation reaction seems to be the formation of the anion **VI/8**, which is resonance stabilized and so more attractive than the alternative anion **VI/5**.

The transamidation reaction can be performed with a large variety of bases, but no systematic investigation has been undertaken. Examples of bases are: potassium 3-aminopropylamide in 1,3-diaminopropane (=KAPA) [3] [4], quinoline [5], NH_3 in a sealed tube [6], KOtBu in different solvents [7], $NaOC_2H_5$ in toluene [7], $KOH/CH_3OH/H_2O$ [7] [8], and KF in dimethylformamide/[18]-crown-6 [9]. Under acidic conditions, *p*-toluolsulfonic acid is a proper catalyst [10].

Using this transamidation reaction, it is possible to introduce three to six ring members in one step, which means that the intermediate ring (see Scheme VI/2) can be five- to eight-membered [6] [7] [11]. The preferred ring size is five to seven; eight-membered intermediates seem to occur only if a β-lactam is used as starting material [6].

The smallest ring which has been enlarged by this method was the four-membered one. Thus, the N-(haloalkyl)-derivatives **VI/11** were prepared from 4-

---

1) Even without any additional reagents **VI/4** converts to its isomer, **VI/9**, after one year in a sealed tube.

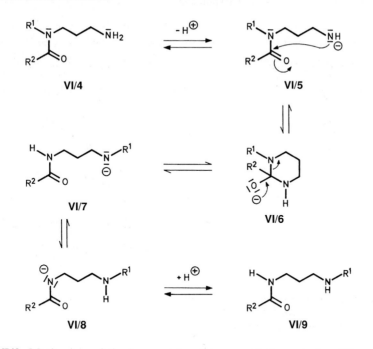

Scheme VI/2. Mechanism of the base catalyzed transamidation reaction [3].

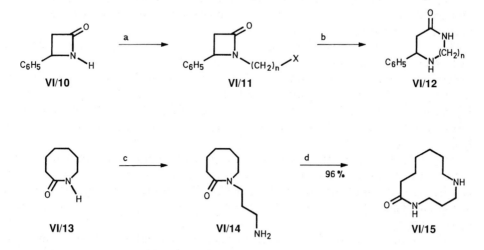

Scheme VI/3. Ring enlargement by transamidation reaction [6] [7].

        X=Cl, Br
        a) X-(CH$_2$)$_n$-X, KOH, THF   b) NH$_3$ liquid
        c) 1. NaOC$_2$H$_5$, CH$_2$=CH-CN 2. H$_2$, Pt, H$_2$SO$_4$, C$_2$H$_5$OH
        d) 1. KAPA, 5 min. 2. H$_3$O$^\oplus$.

phenylazetidin-2-one (**VI/10**) by alkylation with α,ω-dibromoalkane to give the seven- (70 %), eight- (80 %), and nine-membered (67 % yield) azalactams **VI/12**. The enlargement reaction was carried out in liquid NH_3 in a sealed tube (2 to 14 days) Scheme VI/3 [6]. – The treatment of the sodium salt of 7-heptane-lactam (**VI/13**) with acrylonitrile yielded the aminolactam **VI/14** after hydrogenation. In the presence of potassium 3-aminopropylamide in 1,3-diaminopropane (KAPA) the transamidation reaction led to the twelve-membered **VI/15** in 96 % yield (Experimental: as soon as the starting material **VI/14** dissolved completely, the reaction mixture could be worked up). For ring enlargement by three [11] or four atoms, the eight-membered lactam is the smallest which can be used. Large amounts of starting material, and small amounts of 1,8-diazabi-cyclo[5.4.0]undec-7-ene (**VI/19**, =DBU)[2)] were isolated when N-(3-aminopropyl)-ε-caprolactam (**VI/16**) was treated with the KAPA reagent [7]. When the desired eleven-membered azalactam **VI/17** was synthesized by an independent way (electrochemical deprotection of the corresponding tosylated compound **VI/18**), ring contraction (see *e.g.* [12]) took place by a transannular reaction, Scheme VI/4.

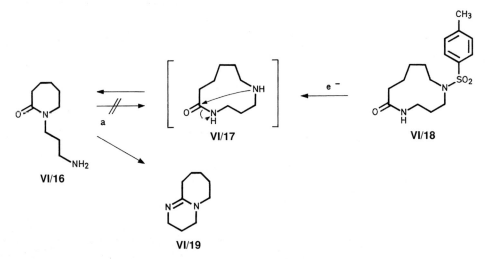

Scheme VI/4. Ring contraction by transamidation reaction [7].
   a) KAPA.

This proved that the seven-membered compound, **VI/16**, with a tertiary lactam nitrogen atom is more stable than the eleven-membered azalactam, **VI/17** with a secondary lactam nitrogen atom. This result seems to be contradictory to the

---

2) Depending on their structural details, bicyclic amidines and amidinium salts hydrolyzed by KOH/H_2O give ring-enlarged or ring-contracted lactams when hydrolyzed by KOH/H_2O [8].

basic mechanism given in Scheme VI/2. However, a comparison of the reactivity of medium sized rings which have two functional groups leads to the conclusion that the eleven-membered ring **VI/17** is less stable than its twelve-membered homologue **VI/15** [13] [14]. Due to the equilibrium between starting material and product, the actual amount of product will depend on the specific reaction conditions. The following conclusions can be drawn from the experiments: The four-membered lactam, a β-lactam, is less stable than the seven- and, interestingly, also the eight-and nine-membered rings; in these cases, the ring enlarged products are stable and no ring contraction properties are reported [6]. Other couples of compounds have been compared: a seven-membered ring is preferred over an eleven; a twelve is preferred over an eight; and a seventeen is preferred over a thirteen.

It is possible to avoid a transannular reaction in a medium-sized ring, *e.g.* a ten-membered one, if the ring is expanded again, Scheme VI/5. Thus, the six-membered barbiturate, **VI/20**, if treated with potassium fluoride, can be enlarged to compound **VI/21** by "Zip reaction" (see later), *via* a ten-membered intermediate [9]. It can be concluded from this reaction that the medium-sized ten-membered ring, although energetically unfavored, will be formed as an intermediate in the presence of non-nucleophilic bases (to avoid lactam cleavage). Probably an isolated ten-membered product will undergo a ring contraction to the starting material or other products.

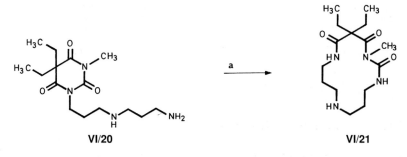

VI/20   VI/21

Scheme VI/5. Ring enlargement of the six-membered lactam **VI/20**, *via* a ten-membered intermediate to **VI/21** [9].

   a) KF, dimethylformamide, [18]crown-6.

Further transamidation reactions are depicted in Scheme VI/6. In some respects, even the Gabriel synthesis, [15] [16] [17][3], a method for preparation of primary amines from N-substituted phthalimides by treatment with hydrazine,

---

3) Transamidation reactions of open chained systems have been known for many years. Treatment of acetamide with aniline leads to acetanilide and ammonia [18]. Similar reactions with hydrazine [19] [20], and with other amines [21] [22] have been described.

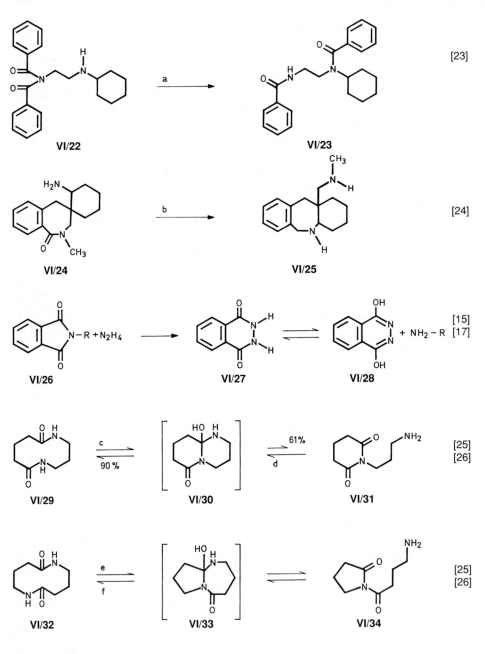

Scheme VI/6. Further examples for transamidation reactions.

R = Alkyl
a) NaOH, H₂O   b) LiAlH₄, THF   c) 0.1 N HCl, H₂O, 20°, 10 d
d) KHCO₃, H₂O   e) 0.1 N HCl, H₂O, 90°, 40 min
f) 0.01 M KHCO₃, H₂O, 1 h.

is an example of transamidation reaction. The driving force in the Gabriel synthesis is the formation of the hydrazide. For analogous transamidation reactions in open chained systems, see ref. [27].

A transannular amide-amide interaction in dilactams has been observed in the reactions of 6,10-dioxo-1,5-diazacyclodecane (**VI/29**) as well as in 5,10-dioxo-1,6-diazacyclodecane (**VI/32**), Scheme VI/6, [25][4]. The N,N'-diacetyl-dilactam derivative of **VI/29** can be obtained in nearly quantitative yield by Beckmann rearrangement of 1,5-cyclooctanedione dioxime ditosylate in dioxane/water in the presence of an excess of sodium acetate. There is one point of interest about this reaction; the rearrangement took place exclusively to give one of the two possible isomers. This is in contrast with other homologous diketones when both isomers are obtained. In order to avoid N-acetyl formation of **VI/29** the rearrangement was carried out in the presence of potassium bicarbonate as base. Compound **VI/29**, left at room temperature in 0.1 N hydrochloric acid, is converted into N-(3-aminopropyl)glutarimide (**VI/31**). The conversion from **VI/31** back to **VI/29** can be carried out by treatment with potassium bicarbonate. In 0.1 N hydrochloric acid, the isomeric dilactam, **VI/32**, gave an equilibrium mixture of **VI/32** and the hydrochloride of N-(4-aminobutyryl)-2-pyrrolidinone (**VI/34**). When this hydrochloride, **VI/34**, was dissolved in 0.01 M bicarbonate (pH 8.75) for one hour, the cyclodipeptide, **VI/32**, could be extracted from the solution. Both reactions probably take place *via* the intermediate cyclols **VI/30** and **VI/33**, respectively; compare ref. [28].

One of the great advantages of the transamidation reaction is that it can be repeated. If the 13-membered lactam, **VI/35**, (R=H) with a 7-amino-4-azaheptyl substituent located at the lactam nitrogen atom is treated with the KAPA reagent, the intermediate **VI/36** should be formed first, see Scheme VI/7.

Compound **VI/36** is in equilibrium with the 17-membered lactam **VI/37**. This compound still contains a tertiary instead of a secondary amide function. Under the same reaction conditions, the substituent at the nitrogen atom can initiate a second transamidation reaction to form the 21-membered lactam **VI/38**. The isomerization of **VI/35** (R=H), in fact, gave 13,17-diaza-20-icosanlactam (**VI/38**, R=H) as the only product in a yield of 90% [30][31]. In the case of **VI/35** (R=CH₃) a mixture of **VI/37** (R=CH₃), and **VI/38** (R=CH₃) was observed (ratio **VI/35**:**VI/37**:**VI/38** approx. 0:1:2) [7]. The same ratio of products was observed when pure **VI/38** was treated under similar conditions. Presumably, the 13-membered **VI/35** still has some transannular strain, which favors the large ring products. – The side chain in **VI/35** can be lengthened by additional couples of propylamine moieties. A treatment of these compounds under the ring enlargement conditions gave 25- [30], 33- [32], and even 53- [33] [34] membered rings. In all three cases the reactions started from a 13-membered precursor, Scheme VI/8.

---

4) A general review of ring chain isomerization is given in ref. [29].

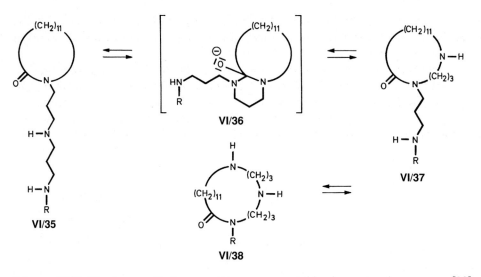

Scheme VI/7. The transamidation reaction as a repeatable ring expansion process [30].

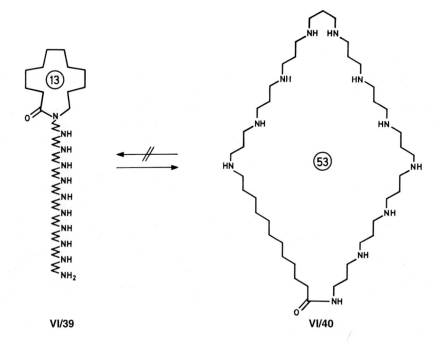

Scheme VI/8. Synthesis of 13,17,21,25,29,33,37,41,45,49-decaaza-52-dopentacontanelactam
(**VI/40**) by Zip reaction from N-(39-amino-4,8,12,16,20,24,28,32,36-nonaaza-
nonatriacontyl)-12-dodecanelactam (**VI/39**) [33].

a) KAPA.

Because the transformation of **VI/39** with a 40-membered ten-amino nitrogen-atom side chain to **VI/40** resembles the unzipping of a sweater, this reaction type has been called the "Zip reaction" [33].

The coumarin derivative, **VI/41**, can be similarly converted into the 14-membered lactam **VI/43** [35]. The first step in this transformation is probably the conjugate addition of a primary amino group of 1,9-diamino-3,7-diazanonane (**VI/42**) to the α,β-unsaturated lactone, **VI/41**, followed by lactone aminolysis and two transamidation reactions (Scheme VI/9). For additional results see ref. [36].

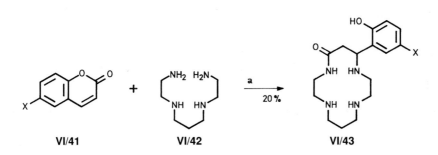

Scheme VI/9. a) CH$_3$OH, 14 d reflux.

The transamidation reaction has been successfully applied to the syntheses of a number of natural polyamino alkaloids. These alkaloids contain principally spermidine or spermine as the basic skeleton [37].

The spermine alkaloid (S)-(-)-homaline (**VI/50**) is the main component of *Homalium pronyense* Guillaum. (Flacourtiaceae). The minor alkaloids from the same plant are, in contrast to **VI/50**, substituted unsymmetrically on the spermine part [38] [39]. The starting material in the first synthesis of (S)-(-)-homaline (**VI/50**) was putrescine (**VI/44**), the central diamino part of the alkaloid, Scheme VI/10, [5]. Tosylation of **VI/45**, prepared by conventional methods from **VI/44**, gave a ditosylate, which reacted in base with (S)-(-)-4-phenyl-2-azetidinone (**VI/46**) to give **VI/47** diprotected with the *tert*-butyloxycarbonyl residue. After **VI/48** was deprotected, the transamidation reaction took place *via* a six-membered intermediate on both ends of the molecule to form the two eight-membered rings in **VI/49**. It is possible that an isomer of **VI/49** with an eleven-membered lactam ring instead of the two eight-membered rings could be formed during the transamidation of **VI/48**. However, because of ring strain differences, the eleven-membered isomer was not observed. The yield (**VI/45** → **VI/50**) was only 7.2 %.

A (-)-homaline synthesis, which is adaptable for the synthesis of the natural unsymmetrical *Homalium* alkaloids, is given in Scheme VI/11 [40] [41]. The eight-membered diaza part, **VI/52**, was synthesized by a transamidation reaction. It was possible to prepare homaline (**VI/50**) and the three other minor bases, hopromine, hopromalinol, and hoprominol, by combination of different eight-membered ring building blocks. As shown in Scheme VI/11, homaline (**VI/50**) was synthesized in 36 % yield.

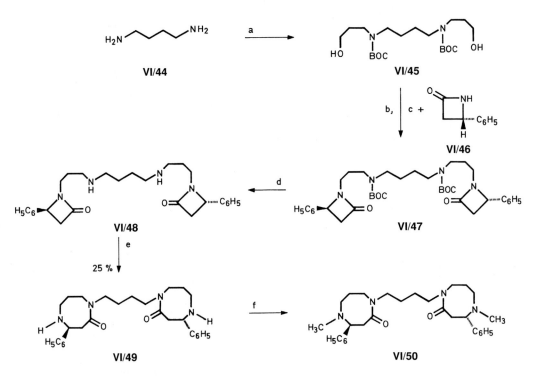

Scheme VI/10. Synthesis of (*S*)-(-)-homaline (**VI/50**) [5].

a) 1. CH$_2$=CH-CN   2. 6 N HCl, H$_2$O   3. C$_2$H$_5$OH, HCl
4. LiAlH$_4$, THF
5. 2-(*tert*-butoxycarbonyloxyimino)-2-phenylacetonitrile   6. (C$_2$H$_5$)$_3$N
b) Ts-Cl, pyridine   c) NaH, dimethylformamide, 98° for 16 h
d) HCOOH   e) (C$_6$H$_5$)$_2$O, Δ   f) CH$_2$O, HCOOH.

(+)-Dihydroperiphylline (**VI/55**, Scheme VI/12), a spermidine alkaloid isolated from *Peripterygia marginata* (Baill.) Loes. [42] was synthesized by a β-lactam-diazacyclooctane method [41] [43]. Thus, treatment of compound **VI/52** with 1-bromo-4-chlorobutane and transformation of the resulting chloro-derivative with liquid ammonia in a sealed tube gave **VI/54**.

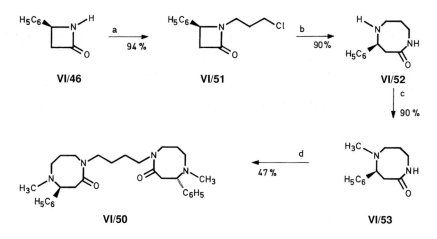

VI/46   **VI/51**   **VI/52**

VI/50   **VI/53**

Scheme VI/11. Synthesis of (S)-(-)-homaline (**VI/50**) [40].

a) Br-(CH$_2$)$_4$-Cl, KOH, THF, Bu$_4$NHSO$_4$ b) NH$_3$, 6 d
c) CH$_2$O, NaBH$_3$CN   d) Br-(CH$_2$)$_4$-Br, KOH, dimethylsulfoxide.

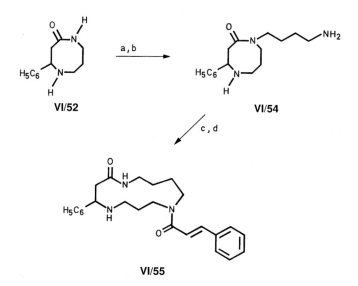

VI/52   **VI/54**

VI/55

Scheme VI/12. Synthesis of (±)-dihydroperiphylline (**VI/55**) [41] [43].

a) Br-(CH$_2$)$_4$-Cl, KN(Si(CH$_3$)$_3$)$_2$, THF, 20°   b) NH$_3$
c) KN(Si(CH$_3$)$_3$)$_2$, THF   d) ClCOCH=CHC$_6$H$_5$, 4-dimethylaminopyridine.

This triamine contains a spermidine backbone, but in contrast to the examples given above for the Zip reaction, the side chain in **VI/54** is a 1,4-diaminobutane. Ring enlargement *via* a seven-membered intermediate must be done with a base of very low nucleophilic character (nucleophilic bases can open the amide functionality, compare *e.g.* [7]). With KN(Si(CH₃)₃)₂ in tetrahydrofurane the 13-membered azalactam **VI/55** was formed in a yield of only 21%.

The alkaloid celacinnine (**VI/61**), isolated from *Maytenus serrata* (Hochst. ex A. Rich.) R. Wilczek and *Tripterygium wilfordii* Hook, is isomeric with dihydroperiphylline (**VI/55**) [44] [45]. Its synthesis is quite remarkable: no less then three different ring enlargement reactions were used to build up the 13-membered ring [46], Scheme VI/13. In the first step, the β-lactam (**VI/46**) was heated with 2-methoxy-pyrroline (**VI/56**). The resulting six-membered **VI/57** gives the nine-membered **VI/58** under the conditions of a reductive cleavage. Specific alkylation at the amide group gives **VI/59**, the starting material for a base catalyzed ring enlargement reaction. Compound **VI/60** was acylated to celacinnine (**VI/61**) in low yield.

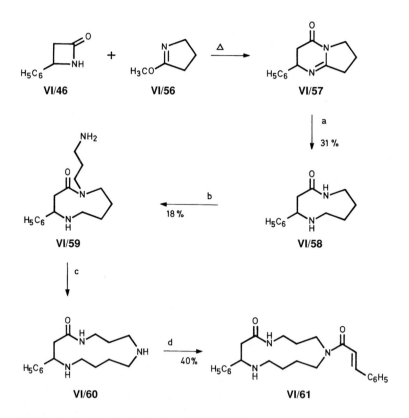

Scheme VI/13. Synthesis of celacinnine (**VI/61**) [46].

a) 1. NaBH₃CN, HOAc, 20°, 24 h 2. NaOH, H₂O,
b) 1. NaH, dimethylformamide, 50° 2. PhthN-(CH₂)₃-I, 25°
3. N₂H₄ · H₂O, C₂H₅OH, reflux
c) 1 N NaOH, 50°   d) (*E*)-C₆H₅CH=CHCOCl.

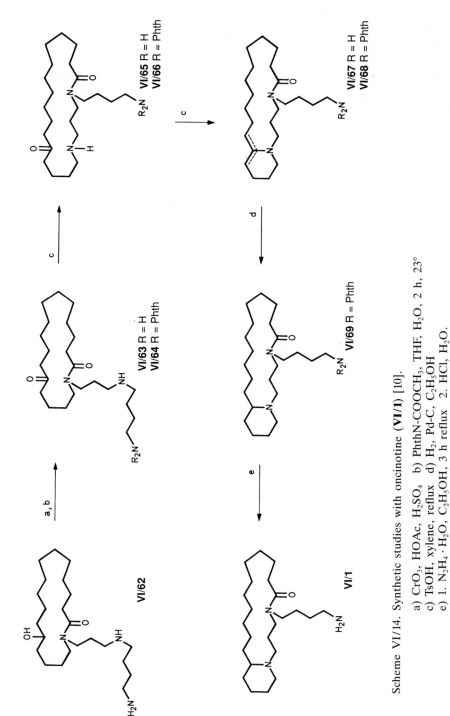

Scheme VI/14. Synthetic studies with oncinotine (**VI/1**) [10].

a) CrO₃, HOAc, H₂SO₄ b) PhthN-COOCH₃, THF, H₂O, 2 h, 23°
c) TsOH, xylene, reflux d) H₂, Pd-C, C₂H₅OH
e) 1. N₂H₄ · H₂O, C₂H₅OH, 3 h reflux 2. HCl, H₂O.

It has been mentioned several times in this chapter (*e.g.* Scheme VI/7) that an equilibrium exists between starting material and product in transamidation reactions. As part of the synthesis of (±)-oncinotine (**VI/1**) it was planned to enlarge compound **VI/63** by four members to get **VI/65** (compare Scheme VI/15). A transannular reaction would follow, and the resulting enamine, **VI/67**, could be catalytically hydrogenated to give oncinotine (**VI/1**), Scheme VI/14.

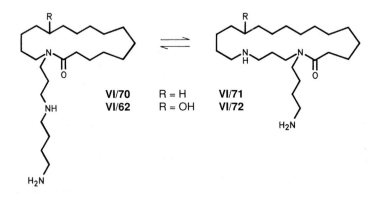

| | VI/70 | R = H | VI/71 |
| | VI/62 | R = OH | VI/72 |

Scheme VI/15. Equilibrium mixtures via transamidation reactions [10] [47] [48].

However, the ketone **VI/63** did not undergo transamidation reaction either under acidic or under basic conditions[5]. When the primary amino function in **VI/63** was protected (as a phthalide by Nefkens reagent [49] [50]), the planned transformation of **VI/64** to **VI/1** *via* **VI/66**, **VI/68**, **VI/69**, outlined above and in Scheme VI/14, was realized in an overall yield of 56 % [10]. The behavior of the ketone **VI/63** suggest an aminoacetal formation of type **VI/73** [51]. Such a compound can be of interest for metallion transport phenomena in plants, because isomers of **VI/63** are natural products and may have some functions in nature. A detailed analysis of this abnormal behavior is in progress [51].

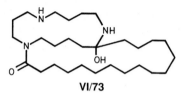

VI/73

---

5) It should be mentioned that a number of other isomers of **VI/63** with the ketone group at different positions behave as **VI/63**: no transamidation reaction takes place.

VI.2. β-Lactams as Synthons for Ring Enlargement

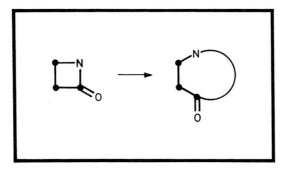

Azetidine-2-ones (β-lactams) are parts of a number of important antibiotic classes such as penicillins [52] [53] [54] [55], cephalosporins [52] [53] [54], and the related classes of thienamycins [52] [56]. During the work on the structures and syntheses of these compounds, many new reactions, including ring enlargement reactions, were observed. However, ring enlargement reactions observed in this field of natural products will not be subject of this chapter. A variety of interesting β-lactam enlargement reactions with more general applications will be described [57].

A number of suitable methods is available for the synthesis of β-lactams and β-lactam derivatives[6], making these compounds good starting materials for further reactions. For mechanistic reasons, we will differentiate these rearrangement reactions into four groups, depending on which ring atom the nucleophile containing substituent is placed. In nearly all examples, the reaction center is the carbonyl group of the lactam.

N-Substituted β-Lactams

If the N-substituent contains a nucleophilic functional group, it is an ideal situation for an attack on the lactam carbonyl group. Thus, acid treatment of 1-(2-aminophenyl)3-ethyl-3-phenyl-azetidine-2-one (**VI/74**[7], Scheme VI/16) gives the diazepine derivative **VI/75** [69] in high yield. If this reaction takes place

---

6) Among the procedures which have been developed for β-lactam syntheses (as reviewed [58] [59]) are: the cyclization of β-halopropionamides [60] from azetidine-2-carboxylic acids and esters [61]; the addition of ketenes to Schiff bases [62]; ring-contraction procedures [63] [64] [65]; [3+1] cyclizations of α-phenylthioacetamide [66]; the cyclization of aminoacids [67], and styrene and N-chlorosulphonyl isocyanate [68].
7) A large variety of similar compounds has been investigated [69].

according to expectation, the configurations at the centres 3 and 4 will not be touched. This can be seen in the conversion, **VI/76** → **VI/77** [57]. Lactone formation can be observed, too, if the nucleophile NH_2 is replaced by OH (**VI/78** → **VI/79** [57], Scheme VI/16).

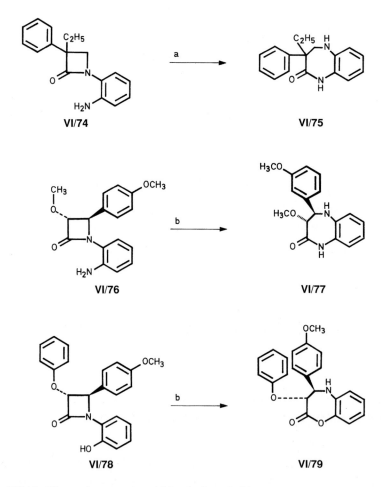

Scheme VI/16. Ring enlargements of N-substituted β-lactams.

a) HCl, H_2O b) organic bases or acids.

During the base catalyzed epimerization of various penicillins, the formation of thiazepinones, **VI/81**, was observed. It has been suggested that the unsaturated β-lactam **VI/80** is formed as a key intermediate by opening of the sulfur containing ring. Nucleophilic attack by the newly freed mercapto group in **VI/80** led to **VI/81** [57], Scheme VI/17.

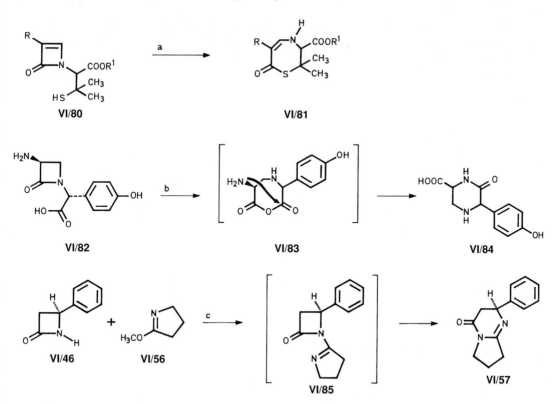

Scheme VI/17. Further examples of rearrangements of N-substituted β-lactams (configuration of **VI/84** not given in [5]).

a) Base   b) silica gel   c) reflux in C_6H_5Cl.

A number of examples of the β-lactam ring enlargement are discussed in Chapter VI.1. Such an enlargement reaction had been used also as a key step in the synthesis of the alkaloid homaline (Scheme VI/10: **VI/48** → **VI/49**; Scheme VI/11: **VI/51** → **VI/52**).

More complex than the examples given above is the formation of the piperazone carboxylic acid **VI/84** from the β-lactam **VI/82**. On silica gel this reaction takes place slowly. A plausible explanation is given in Scheme VI/17 [5]. The rearrangement takes place *via* an intramolecular attack of the neighboring carboxylic acid residue to give the seven-membered cyclic anhydride, **VI/83**. Because of the primary nature of one of the amino groups a second rearrangement, an internal acylation, can then take place. It should be noted that the second rearrangement is a ring contraction to give a stable six-membered ring.

Finally there is a reaction of a N-substituted β-lactam which had been used several times in the syntheses of natural occurring polyamine alkaloids [70]: the β-lactam **VI/46** reacts on heating with 2-methoxypyrroline **VI/56** [71] to form the dihydropyrimidine derivative **VI/57**[8]. The proposed intermediate **VI/85**, given in Scheme VI/17, has not been isolated [72]. The yield seems to be depend very much on the structure of the methyl imino ether. Presumably, geometric effects associated with the *syn* versus *anti* configurations of the cyclic imino ether groups in medium and large ring compounds may affect the course of this reaction. The whole mechanism is as yet unknown, but analogues of the inter-mediate **VI/85** have been synthesized [70]:

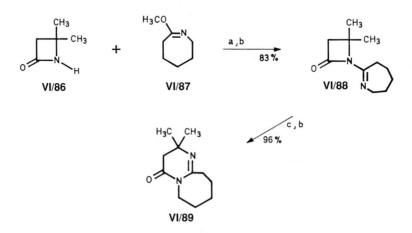

Scheme VI/18. a) 130°, 2 h   b) without solvent   c) 180°, 2 h.

In contrast to **VI/85** compound **VI/88** (Scheme VI/18) is a stable, crystalline, colorless material. It undergoes base catalyzed conversion to **VI/89** [70]. This reaction has been applied as a key step in the synthesis of the following natural products : celacinnine [46] (compare Chapter VI.1), O-methylorantine [74], chaenorhine [75], cannabisativine [76], and verbascenine [72].

β-Lactams Substituted at Position 3

As shown above, the ring enlargement of N-substituted β-lactams gives com-pounds in which the lactam ring is enlarged by *n* members. The letter *n* stands for the number of atoms placed between the β-lactam ring and the nucleophile. In case of compound **VI/80** (Scheme VI/17) the new ring should be seven-mem-bered: 4 (β-lactam) + 3 (three-membered side chain) = 7. This kind of calcula-tion will give incorrect results if applied to β-lactams substituted at position 3.

---

8) For other methods to synthesize dihydropyrimidines of type **VI/57** see ref. [72] [73].

The nucleophile in the side chain will attack the carbonyl group of the lactam and, in the course of the ring enlargement, the atom 1 acts as an internal leaving group. The first example in Scheme VI/19 is tabtoxin (**VI/90**).

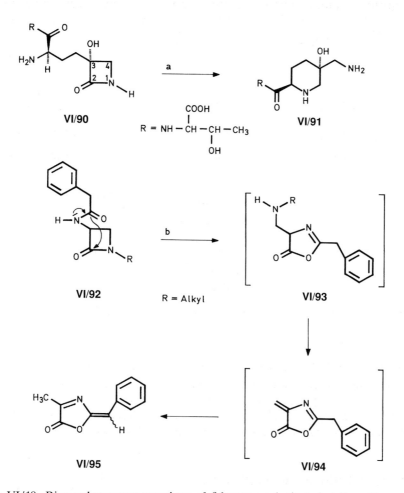

Scheme VI/19. Ring enlargement reactions of β-lactams substituted at 3-position.
a) 20° b) reflux in C₆H₅OCH₃.

This compound undergoes a translactamization reaction to give the non-toxic isotabtoxin (**VI/91**) at room temperature and neutral pH with a half-life of about one day [77] [78] [79]. Inactivation of tabtoxin is even quicker in either acid or base [80]. (The driving force of this conversion seems to be the change to a less strained molecule). The transformation of the β-lactam to the five-membered **VI/95** (by heating in anisol) is another example of this type and includes the ring enlargement to **VI/93** followed by an elimination reaction to give **VI/94** [57]. – Finally, an attempt was made to prepare the amino alcohol, **VI/97**, by condensation of the β-lactam, **VI/96**, and ethyleneoxide. However, the only reaction product found, was the morpholine **VI/98** [57], Scheme VI/20. Compound **VI/98** is explained as a rearranged product of the first formed **VI/97**.

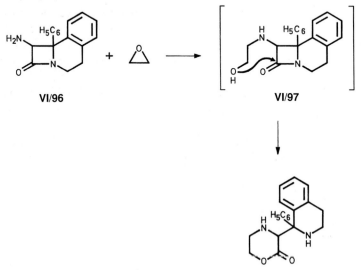

Scheme VI/20. β-Lactam ring expansion.

β-Lactams Substituted at Position 4

If a nucleophilic group is placed in a side chain which is located at β-lactam position 4, as shown in Scheme VI/21, the size of the new ring formed be six. The transformation from **VI/99** to **VI/100** takes place in acid or base [57]. The analogous formation of **VI/102** has been observed when the trimethylsilyl protected β-lactam **VI/101** was deprotected [81]. In both cases, retention (C(3)) and inversion (C(4)) of the configuration was found.

Other Types of β-Lactam Rearrangements

The lactam bond can be opened photochemically in an α-fission. If an appropriate substituent is located at the nitrogen atom the radical stabilization can take place as shown in Scheme VI/22, **VI/103** $\rightarrow$ **VI/106** [82] [83][9]. Dihydrouracils of

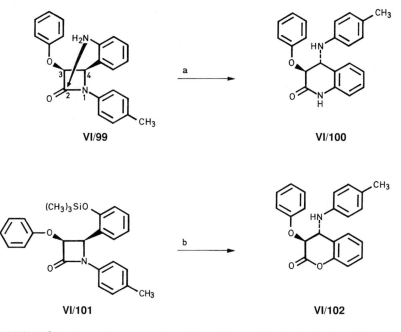

Scheme VI/21. β-Lactam ring expansion effected by a side chain in 4-position.
a) TsOH or N(C₂H₅)₃   b) CH₃OH.

type **VI/108** are formed if the enamine $(CH_3)_2C=CHN(CH_3)_2$ prepared from dimethylamine and 2-methylpropanal, is heated with isocyanates in a sealed tube at about 120°. It is likely that a β-lactam (compare **VI/107**) reacts again as an intermediate. In the presence of an excess of isocyanate a second molecule of isocyanate will be involved in this reaction [84] [85] [86]. – Attempts have been made to synthesize β-thiolactam analogues of penam and cepham systems. In fact 1:1 adducts were obtained from silylthioketenes such as **VI/109** and 2-thiazolines of type **VI/110**, Scheme VI/22, but these products proved to be

---

9) The photochemical ring enlargement of N-phenyl-lactams does not work with four-, five-, and six-membered lactams [82]. The reaction has been observed only in the case of benzannelated β-lactams, see Scheme VI/22 [83]. Photolysis of larger N-phenyl-lactams are as follows [82]:

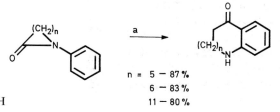

n = 5 — 87 %
      6 — 83 %
     11 — 80 %

a) hν, C₂H₅OH

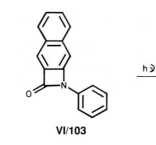

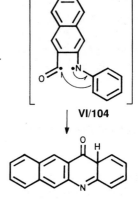

VI/103

VI/104

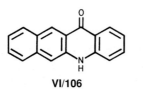

VI/106

VI/105

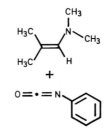

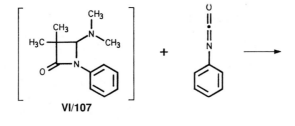

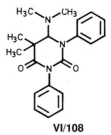

VI/107

VI/108

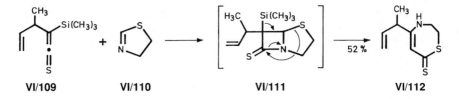

VI/109   **VI/110**   **VI/111**   52 %   **VI/112**

Scheme VI/22. Other types of β-lactam ring expansions.

thiazepine derivatives **VI/112**. They result from rearrangement of an intermediate thiopenam system, **VI/111** [87].

β-Lactams participate in a number of ring enlargement reactions of different types. Because of the classification in this review they are described at other places too; *e.g.* Sommelet-Hauser rearrangement (Chapter V.3).

VI.3. Cyclodepsipeptides

Cyclodepsipeptides are natural products and one of the methods for their syntheses includes a ring enlargement approach. The two reactions mentioned in Scheme VI/23 are examples. Neighboring hydroxy-amide interactions in

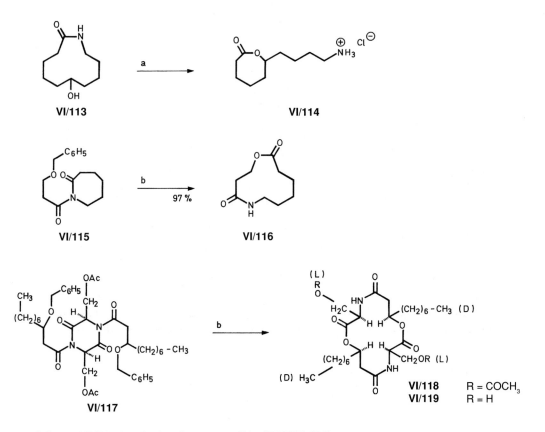

Scheme VI/23. Synthesis of serratamolide (**VI/119**) [89].

a) HCl, H₂O   b) H₂, Pd-black, THF.

lactams have been noticed [88], when, for example, the eleven-membered 6-hydroxydecane-10-lactam (**VI/113**) was converted to the 6-(4-aminobutyl)-hexane-6-lactone (**VI/114**)[10] under acid catalysis. A similar but, concerning the ring size, reverse reaction leading to the eleven-membered cyclodepsipeptide **VI/116** took place spontaneously when the benzyloxy compound **VI/115** was hydrogenated [89].

In the first reaction (**VI/113** → **VI/114**) a secondary amide (lactam) was transformed to a lactone and a primary amide. Presumably the conversion is possible only under acidic conditions, because, in the presence of base, the resonance stabilized secondary lactam will have been formed, and this will stabilize the medium-sized eleven-membered ring. The ring strain in the five-, six-, and seven-membered rings is less than in the ten- and eleven-membered ones. Under basic or neutral reaction conditions larger rings are preferred to the smaller rings, because of the resonance stabilized secondary lactams. The high yield (97 %) of the eleven-membered compound **VI/116** was observed under neutral conditions. This shows that the seven-membered imide with a primary alcoholic group, prepared from **VI/115**, is less stable than the medium-sized **VI/116**, containing a secondary lactam and a lactone group.

The antibiotic serratamolide (**VI/119**, Scheme VI/23), a naturally occurring cyclodepsipeptide, was isolated from a *Serratia marcescens* culture [90]. Its synthesis is an application of the principal method described above. The hydrogenolysis of the diketopiperazine **VI/117** gives the O,O'-diacetyl derivative **VI/118**, which was converted into the antibiotic itself by mild hydrolysis [91].

A general procedure for the synthesis of cyclic depsipeptides was published recently [92]. Starting material is the open chained compound of structure **VI/120**, Scheme VI/24.

It can be prepared by treatment of 3-amino-2*H*-azirines (*e.g.* 3-(dimethyl-amino)-2,2-dimethyl-2*H*-azirine) with an amino acid or peptide and, finally, with a ω-hydroxyacid. The formation of the oxazolone, **VI/121**, is observed when **VI/120** is treated with acid. The ring enlargement step, the conversion of **VI/121** to **VI/122**, is observed under the same conditions. The transformation of (-)-(R,R)2-{2-[2-(2-hydroxy-2-phenylacetamido)-2-methylpropionato]-2-phenylacetamido}-N,N,2-trimethylpropionamide (**VI/123**) to (-)-(R,R)-3,3,9,9-tetramethyl-6,12-diphenyl-1,7-dioxa-4,10-diazacyclododecane-2,5,8,11-tetrone (**VI/126**) in dry toluene/hydrochloric acid at 100° was observed in a 88 % yield. Compounds **VI/124** and **VI/125** are discussed to be intermediates. In an analogous reaction sequence cyclopeptides can be synthesized [93].

---

10) Experiments for the reverse reaction **VI/114** → **VI/113** have not been described, ref. [88].

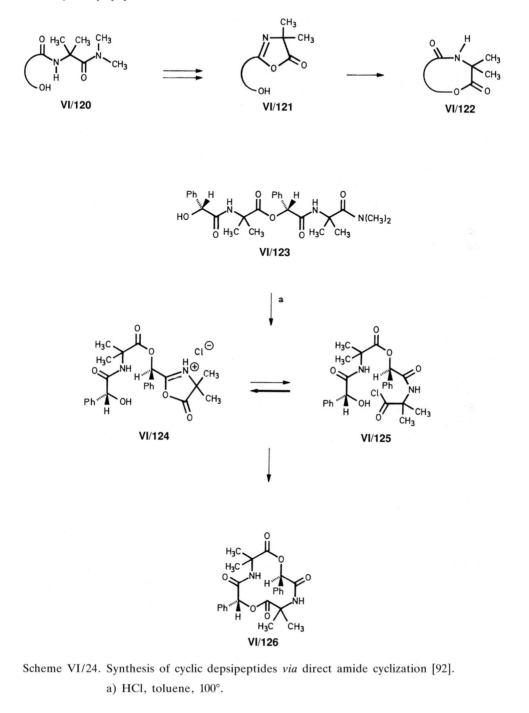

Scheme VI/24. Synthesis of cyclic depsipeptides *via* direct amide cyclization [92].

a) HCl, toluene, 100°.

References

[1] M. M. Badawi, A. Guggisberg, P. v. d. Broek, M. Hesse, H. Schmid, Helv.Chim.Acta **51**, 1813 (1968).

[2] A. Guggisberg, M. M. Badawi, M. Hesse, H. Schmid, Helv.Chim.Acta **57**, 414 (1974).

[3] A. Guggisberg, B. Dabrowski, U. Kramer, C. Heidelberger, M. Hesse, H. Schmid, Helv.Chim.Acta **61**, 1039 (1978).

[4] C. A. Brown, J.Chem.Soc., Chem.Commun. **1975**, 222.

[5] H. H. Wasserman, G. D. Berger, K. R. Cho, Tetrahedron Lett. **23**, 465 (1982).

[6] L. Crombie, R.C. F. Jones, S. Osborne, A. R. Mat-Zin, J.Chem.Soc., Chem.Commun. **1983**, 959.

[7] A. Guggisberg, U. Kramer, C. Heidelberger, R. Charubala, E. Stephanou, M. Hesse, H. Schmid, Helv.Chim.Acta **61**, 1050 (1978).

[8] C. Heidelberger, A. Guggisberg, E. Stephanou, M. Hesse, Helv.Chim.Acta **64**, 399 (1981).

[9] C. Jenny, M. Hesse, Helv.Chim.Acta **64**, 1807 (1981).

[10] S. Bienz, A. Guggisberg, R. Wälchli, M. Hesse, Helv.Chim.Acta **71**, 1708 (1988).

[11] E. Stephanou, A. Guggisberg, M. Hesse, Helv.Chim.Acta **62**, 1932 (1979).

[12] N. J. Leonard, Rec.Chem.Progr. **17**, 243 (1956).

[13] A. C. Cope, M. M. Martin, M. A. McKervey, Quart.Rev. **20**, 119 (1966).

[14] G. Haufe, M. Mühlstädt, Z.Chem. **19**, 170 (1979).

[15] M. S. Gibson, R.W. Bradshaw, Angew.Chem. **80**, 986 (1968), Angew.Chem.Int. Engl.Ed. **7**, 919 (1968).

[16] C. C. DeWitt, Org.Synth., Coll.Vol. **II**, 25 (1943).

[17] L. H. Amundsen, J. J. Sanderson, Org.Synth., Coll.Vol. **III**, 256 (1955).

[18] W. Kelbe, Ber.dtsch.chem.Ges. **16**, 1199 (1883).

[19] F. Just, Ber.dtsch.chem.Ges. **19**, 1201 (1886).

[20] M. Freund, B. B. Goldsmith, Ber.dtsch.chem.Ges. **21**, 2456 (1888).

[21] A. Galat, G. Elion, J.Am.Chem.Soc. **65**, 1566 (1943).

[22] H. R. Hirst, J. B. Cohen, J.Chem.Soc. **67**, 829 (1895).

[23] C. J. M. Stirling, J.Chem.Soc. **1958**, 4531.

[24] R. Süess, Helv.Chim.Acta **62**, 1103 (1979).

[25] G. I. Glover, R. B. Smith, H. Rapoport, J.Am.Chem.Soc. **87**, 2003 (1965).

[26] S. Wolfe, S. K. Hasan, Can.J.Chem. **48**, 3566 (1970).

[27] T. Wieland, H. Urbach, Liebig Ann.Chem. **613**, 84 (1958).

[28] V. K. Antonov, T.E.Agadzhanyan, T. R. Telesnina, M. M. Shemyakin, G. G. Dvoryant-seva, Y. N. Sheinker, Tetrahedron Lett. **1964**, 727.

[29] R. E. Valter, "Ring chain isomery in organic chemistry", Sinatne, Riga, 1978.

[30] U. Kramer, A. Guggisberg, M. Hesse, H. Schmid, Helv.Chim.Acta **61**, 1342 (1978).

[31] U. Kramer, A. Guggisberg, M. Hesse, H. Schmid, Angew.Chem. **89**, 899 (1977), Angew.Chem.Int.Ed.Engl. **16**, 861 (1977).

[32] U. Kramer, H. Schmid, A. Guggisberg, M. Hesse, Helv.Chim.Acta **62**, 811 (1979).

[33] U. Kramer, A. Guggisberg, M. Hesse, H. Schmid, Angew.Chem. **90**, 210 (1978), Angew.Chem.Int.Ed.Engl. **17**, 200 (1978).

[34] U. Kramer, A. Guggisberg, M. Hesse, Helv.Chim.Acta **62**, 2317 (1979).

[35] E. Kimura, T. Koike, M. Takahashi, J.Chem.Soc., Chem.Commun. **1985**, 385.

[36] E. Kimura, T. Koike, K. Uenishi, M. Hediger, M. Kuramoto, S. Joko, Y. Arai, M. Kodama, Y. Iitaka, Inorg.Chem. **26**, 2975 (1987).

[37] A. Guggisberg, M. Hesse, in "The Alkaloids" (Ed. A. Brossi) **22**, 85 (1983).

[38] M. Pais, R. Sarfati, F.-X. Jarreau, R. Goutarel, Tetrahedron **29**, 1001 (1973).

[39] M. Pais, R. Sarfati, F.-X. Jarreau, R. Goutarel, Comp.rend.Acad.Sci. Paris **C 272**, 1728 (1971).

[40] L. Crombie, R. C. F. Jones, A. R. Mat-Zin, S. Osborne, J.Chem.Soc., Chem.Commun. **1983**, 960.

[41] L. Crombie, R. C. F. Jones, D. Haigh, Tetrahedron Lett. **27**, 5147 (1986).

[42] R. Hocquemiller, A. Cavé, H.-P. Husson, Tetrahedron **33**, 645 (1977).

[43] L. Crombie, R. C. F. Jones, D. Haigh, Tetrahedron Lett. **27**, 5151 (1986).

[44] S. M. Kupchan, H. P. J. Hintz, R. M. Smith, A. Karim, M.W. Cass, W. A. Court, M. Yatagai, J.Chem.Soc., Chem. Commun. **1974**, 329.

[45] S. M. Kupchan, H. P. J. Hintz, R. M. Smith, A. Karim, M.W. Cass, W. A. Court, M. Yatagai, J.Org.Chem. **42**, 3660 (1977).

[46] H. H. Wasserman, R. P. Robinson, H. Matsuyama, Tetrahedron Lett. **21**, 3493 (1980).

[47] R. Wälchli, A. Guggisberg, M. Hesse, Helv.Chim.Acta **67**, 2178 (1984).

[48] R. Wälchli, A. Guggisberg, M. Hesse, Tetrahedron Lett. **25**, 2205 (1984).

[49] G.H. L. Nefkens, G. I. Tesser, R. J. F. Nivard, Rec.Trav.Chim. Pays-Bas **79**, 688 (1960).

[50] P. M. Worster, C. C. Leznoff, C. R. McArthur, J.Org.Chem. **45** 174 (1980).

[51] H. Benz, A. Guggisberg, M. Hesse, 1990 to be published.

[52] K. Nakanishi, T. Goto, S. Itô, S. Natori, S. Nozoe (eds.) Natural Products Chemistry Vol. 3, Oxford University Press, Oxford 1983.

[53] W. Dürckheimer, F. Adam, G. Fischer, R. Kirrstetter, Adv.Drug Res. **17**, 61 (1988).

[54] T. F. Walsh, Ann.Rep.Med.Chem. **23**, 121 (1988).

[55] E. H. Flynn, "Cephalosporins and Penicillins" (Ed.), Academic Press, New York 1972.

[56] D.B. R. Johnston, S. M. Schmitt, F. A. Bouffard, B. G. Christensen, J.Am.Chem.Soc. **100**, 313 (1978).

[57] M. S. Manhas, S. G. Amin, A. K. Bose, Heterocycles **5**, 669 (1976).

[58] A. K. Mukerjee, R. C. Srivastava, Synthesis **1973**, 327.

[59] N. S. Isaacs, Chem.Rev. **76**, 181 (1976).

[60] H. H. Wasserman, D. J. Hlasta, A.W. Tremper, J. S. Wu, Tetrahedron Lett. **1979**, 549.

[61] H. H. Wasserman, B. H. Lipshutz, A.W. Tremper, J. S. Wu, J.Org.Chem. **46**, 2991 (1981).

[62] A. K. Bose, H. P. S. Chawla, B. Dayal, M. S. Manhas, Tetrahedron Lett. **1973**, 2503.

[63] G. Lowe, D. D. Ridley, J.Chem.Soc. Perkin Trans. I **1973**, 2024.

[64] G. Stork, R. Szajewski, J.Am.Chem.Soc. **96**, 5787 (1974).

[65] D. Bender, H. Rapoport, J. Bordner, J.Org.Chem. **40**, 3208 (1975).

[66] K. Hirai, Y. Iwano, Tetrahedron Lett. **1979**, 2031.

[67] L. Birkhofer, J. Schramm, Liebigs Ann.Chem. **1975**, 2195.

[68] R. Graf, Liebigs Ann.Chem. **661**, 111 (1963).

[69] B. J. R. Nicolaus, E. Bellasio, G. Pagani, L. Mariani, E. Testa, Helv.Chim.Acta **48**, 1867 (1965).

[70] D. Bormann, Chem.Ber. **103**, 1797 (1970).

[71] H. H. Wasserman, R. P. Robinson, H. Matsuyama, Tetrahedron Lett. **21**, 3493 (1980).

[72] S. Petersen, E. Tietze, Liebigs Ann.Chem. **623**, 166 (1959).

[72] H. H. Wasserman, R. P. Robinson, Tetrahedron Lett. **24**, 3669 (1983).

[73] E. Profft, F.-J. Becker, J.prakt.Chemie **30**, 18 (1965).

[74] H. H. Wasserman, R. K. Brunner, J. D. Buynak, C. C. Carter, T. Oku, R. P. Robinson, J.Am.Chem.Soc. **107**, 519 (1985).

[75] H. H. Wasserman, R. P. Robinson, C. G. Carter, J.Am.Chem.Soc. **105**, 1697 (1983).

[76] H. H. Wasserman, M. R. Leadbetter, Tetrahedron Lett. **26**, 2241 (1985).

[77] D. L. Lee, H. Rapoport, J.Org.Chem. **40**, 3491 (1975).

[78] D. L. Selwood, K. S. Jandu, Trop.Med.Parasitol. **39**, 81 (1988).

[79] P. A. Taylor, H. K. Schnoes, R. D. Durbin, Biochem.Biophys.Acta **286**, 107 (1972).

[80] W.W. Stewart, Nature **229**, 174 (1971).

[81] A. K. Bose, S. D. Sharma, J. C. Kapur, M. S. Manhas, Synthesis **1973**, 216.

[82] M. Fischer, Chem.Ber. **102**, 342 (1969).

[83] G. Ege, E. Beisiegel, Angew.Chem. **80**, 316 (1968); Angew.Chem.Int.Engl.Ed. **7**, 303 (1968).

[84] G. Opitz, J. Koch, Angew.Chem. **75**, 167 (1963); Angew.Chem.Int.Engl.Ed. **2**, 152 (1963).

[85] M. Perelman, S. A. Mizsak, J.Am.Chem.Soc. **84**, 4988 (1962).

[86] A. K. Bose, G. Mina, J.Org.Chem. **30**, 812 (1965).

[87] J. Lindstaedt cited in E. Schaumann, Tetrahedron **44**, 1827 (1988).

[88] J. A. Davies, C. H. Hassall, I. H. Rogers, J.Chem.Soc. C **1969**, 1358.

[89] M. M. Shemyakin, V. K. Antonov, A. M. Shkrob, V. I. Shchelokov, Z. E. Agadzhanyan, Tetrahedron **21**, 3537 (1965).

[90] H. H. Wasserman, J. J. Keggi, J. E. McKeon, J.Am.Chem.Soc. **84**, 2978 (1962).

[91] M. M. Shemyakin, Y. A. Ovchinnikov, V. K. Antonov, A. A. Kiryushkin, V. T. Ivanov, V. I. Shchelokov, A. M. Shkrob, Tetrahedron Lett. **1964**, 47.

[92] D. Obrecht, H.Heimgartner, Helv.Chim.Acta **73**, 221 (1990).

[93] J. M. Villalgordo, T. Linden, H. Heimgartner, Helv.Chim.Acta 1990 to be published.

VII. Ring Enlargement by Side Chain Incorporation

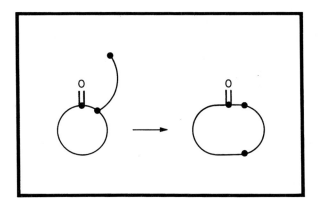

This chapter presents a series of reactions of cycloalkanones disubstituted in position 2. One of the substituents is a side chain of varying length, containig, at its end, the residue X. This residue X represents an internal nucleophile, which attacks the carbonyl carbon atom, in order to form a bicyclic alcohol intermediate. Oxygen, nitrogen, or carbon nucleophiles can be used. The second substituent in cycloalkanone position 2 is Y, representing an electron withdrawing group, such as COR, SO_2R, NO_2, CN, NO [1], and COOR. Such a group allows the ring enlarging step. This consists of a heterolytic cleavage of the bridge bond in the bicyclic intermediate. The splitting of the same bond can also be observed, if homolytic reaction conditions are used. In the latter case the residue Y represents an alkyl or COOR group.

There are several known examples, in which bicyclic intermediates, outlined above, were isolated and yielded the expected ring enlargement products on further treatment with base. In order to study the stereochemical course of the enlargement reaction, the relative configurations of the bicycles **VII/1** [2], **VII/3** [3], and **VII/5** [4] (Scheme VII/1) were determined by X-ray structure analysis. Intermediates **VII/1** and **VII/3** are *trans-*, and **VII/5** is *cis-* fused. Both **VII/3** and **VII/5**, in spite of having different configurations, under basic conditions, undergo ring expansion to **VII/4** and **VII/6**, respectively. While the yields in the second and third examples are high, the conversion rate of **VII/1**→ **VII/2** is distinctly lower, Scheme VII/1. Assuming all three examples proceed through the same reaction mechanism and therefore have the same stereochemical arrangement at the bicyclic bridge, it must be concluded that one or even two of the diastereoisomers are not the direct percursors of the corresponding ring expanded products. In such cases, base-catalyzed epimerisation of the hemi-acetal functions must take place. This epimerization can be brought about by a retro-aldol reaction, to generate the apropriate diastereoisomer, to undergo ring expansion[1]. No other stereochemical details are known in the mechanistic course of the heterolytic expansion reaction.

This Chapter is subdivided into carbocycle, lactam, and lactone forming reactions according to the structural classes of the products formed by the ring enlargement.

---

1) An isomerization prior to the ring expansion has been observed. Thus, aldol product **A** containing a six-membered ring gave the ring enlarged product **B**, which can only be derived from the alternative six-membered intermediate, **C** [5].

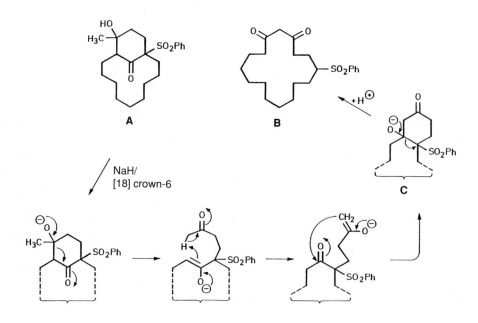

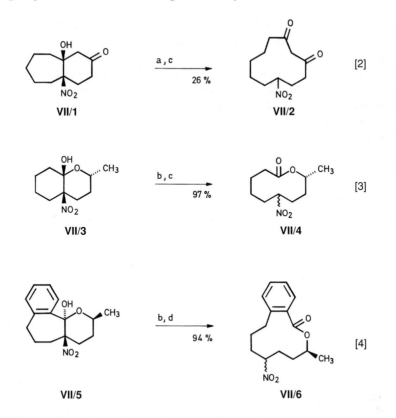

Scheme VII/1. Configurations of bicyclic intermediates of the ring enlargement reaction.
a) KOtBu, THF   b) cat. Bu$_4$NF, THF   c) HOAc   d) H$_2$O.

VII.1. Ring Expansion Reactions Leading to Carbocycles

The incorporation of the side chain, placed in 2-position of a cycloalkanone, is influenced by strong electron withdrawing groups as Y. There are a number of experiments published, in which the nitro, the phenylsulfonyl, the ketone carbonyl, and the cyano group were used as Y.

The heterolytic cleavage of the C(1),C(2) bond of 2-nitroketones, substituted at the 2-position has been investigated extensively, and the synthesis of 2-nitroketones and their reactions with external nucleophiles have been reviewed comprehensively [6]. The activated C(1),C(2) bond in 2-nitrocycloalkanones, alkylated in position 2, can be cleaved with an external nucleophile [7]. The resulting open chain compounds represent useful building blocks in organic synthesis [8] [9]. For our purposes, cleavage of the C(1),C(2) bond with internal C-nucleophiles can afford ring enlarged products.

Open chain 2-nitroalkanones substituted by a second alkanone residue in the 2-position are the first examples which will be discussed, Scheme VII/2. 2-Methyl-2-nitro-1-phenyl-1,5-hexanedione (**VII/7**), prepared from 2-nitro-1-phenyl-1-propanone and methylvinylketone in the presence of triphenylphosphine, as well as its homologue, **VII/10**, under base catalysis, rearrange to the 1,3-diketones **VII/9** and **VII/12**, respectively. In contrast to the starting compounds, **VII/7** and **VII/10**, both products (**VII/9** and **VII/12**) in basic medium can form resonance stabilized dianions (1,3-diketone and secondary nitroalkane moiety). This behavior seems to be the driving force for the rearrangement. It is remarkable that both homologues **VII/7** and **VII/10** behave differently under the same conditions: the methylketone **VII/7** rearranges *via* the six-membered intermediate **VII/8**, and the ethylketone **VII/10** goes exclusively *via* the four-membered **VII/11**. In both cases the alternative products have not been detected.

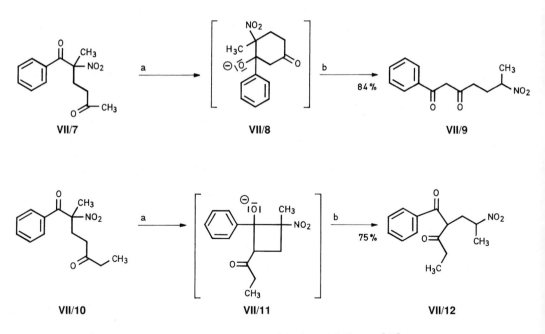

Scheme VII/2. Base catalyzed rearrangement of 2-nitro-1,5-diones [10].
a) KOtBu, THF, −78° b) HOAc.

In addition to the two alternative active methylene groups in the side chain ring, strain effects have to be considered when cyclic analogues of **VII/7** and **VII/10** are treated with base. Thus, compounds with the general formula **VII/13** give, apart from some side products (*e.g.* retro-Michael and bicyclic compounds

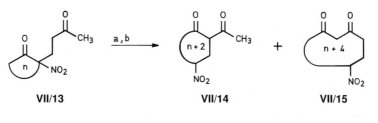

	Ratio		Yield
n	VII/14	VII/15	
6	100	0	85%
7	56	44	79%
8	11	89	45%

Scheme VII/3. Rearrangement of cyclic 2-nitro-1,5-diones [2].

n = Ring size
a) 2 KOtBu, THF, 80° b) $H_3O^{\oplus}$.

such as **VII/1**), a mixture of **VII/14** (incorporation of two carbon atoms into the ring) and **VII/15** (four carbon atoms), see Scheme VII/3. Several conclusions can be drawn from these results : The eight-membered ring is preferred to the ten-membered ring; nine- and eleven-membered ring compounds such as **VII/14** and **VII/15** have comparable stabilities; and, finally, in comparing ten- and twelve-membered rings, the latter is preferred. Steric properties of the dianions, **VII/14** and **VII/15** may also be responsible for the different product ratios. Another conclusion involves the acidity of the active methylene group situated in the side chain. This was studied using a side chain containing a β-ketoester moiety. In this case, the ring expansion took place regiospecifically and in high yields with insertion of four carbon atoms, as shown in Scheme VII/4[2].

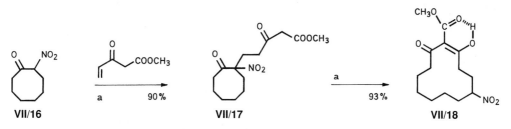

Scheme VII/4. Regiospecific rearrangement of 2-nitro-1,5-dione **VII/17** [11].

a) Bu_4NF/THF.

---

2) The twelve-membered **VII/18** in $CHCl_3$ solution exists in four conformers (NMR spectra). Most similar compounds also have more than one conformation. – Hydrolysis of the ester in **VII/18** was not possible, and decarboxylation could only be realized by hydrogenolyses of the corresponding benzylester [2].

The smallest ring expanded by this method was seven-membered (**VII/13**, n = 6) [2] [4] [11]. The number of carbon atoms incorporated by this method ranges between two and four. A side reaction predominates when attempts were made to incorporate a three carbon atom unit from a butanoate chain as in **VII/19**. Instead of the expected ring enlargement product, a nitrone was observed, formed by a direct attack of the carbanion at the nitro nitrogen atom, Scheme VII/5 [12].

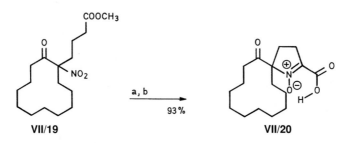

Scheme VII/5. Nitrone formation as a side reaction [12].

a) KOtBu, THF   b) H$_2$O.

In situ generated enamines[3] from compound **VII/22**, prepared from 2-nitro-cycloalkanones, undergo ring expansion reactions too. The conditions are particularly mild, and the yields are high. This reaction was applied to the syntheses of cyclopentadecanone (**VII/25**, R=H) and (±)-muscone (**VII/25**, R=CH$_3$), given in Scheme VII/6 [14]. The preparation of the aldehyde, **VII/22**, from 2-nitrocyclododecanone (**VII/21**) was initiated by a Pd(O)-mediated alkylation, followed by functional group modification reactions. The ring enlargement step took place in alcoholic solution at room temperature in the presence of pentylamine. Denitration with tributyltinhydride was the low yield step in this synthesis.

The phenylsulfonyl residue is another functional group (Y) with high electron withdrawing capacity. Different authors have used it for ring enlargement reactions [5] [15] [16]. One of the procedures is the following: During the ring expansion step a three carbon atom unit will be incorporated into the smaller ring [15]. The transformations of a five- to eight-, and an eight- to eleven-membered rings will result. The three carbon atom expansion seems to be the ideal way to transform the inexpensive cyclododecanone into the expensive 15-membered muscone (**VII/25**, R=CH$_3$). Such a synthesis has been carried out in the following way: Phenylsulfonyl substitution in position 2 of cyclododecanone

---

3) Pentylamine gave the best results in the enamine ring enlargement route. For comparable reactions, see ref. [13].

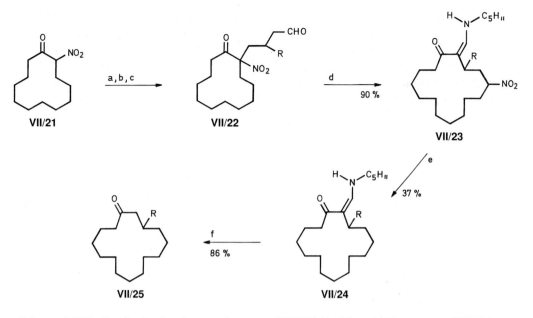

Scheme VII/6. Synthesis of cyclopentadecanone (**VII/25**, R=H) and (±)-muscone (**VII/25**,
R=CH₃) by the enamine route of ring enlargement (yields are given for
R=H) [14].

a) CH₂=CR-CH(COOCH₃)(OCOOC₂H₅), Pd[C₆H₅)₃P]₄ b) H₂, Pd-C
c) diisobutylaluminiumhydride   d) H₂N-C₅H₁₁, C₂H₅OH
e) Bu₃SnH, 2,2'-azobisisobutyronitrile, toluene
f) KOH, C₂H₅OH, H₂O, 15 h, reflux.

takes place by bromination (Br₂, CHCl₃, 20°). Followed by treatment with
sodium phenylsulfinate (dimethylformamide, 120°) to give **VII/26**. Reaction of
the mesylate, **VII/27**, with the sodium enolate of the β-keto sulfone **VII/26**
(Scheme VII/7) in the presence of sodium iodide in dimethylformamide affords
the desired alkylated product, **VII/28**. Treatment of this compound with a
catalytic amount of fluoride ion (Bu₄NF) leads directly to the ring-expansion
product, **VII/30**. No intermediates can be detected. However, intermediates
with a *cis*-substitution in the bicyclic bridge can be isolated in cases of model
compounds with smaller rings. Presumably, the reaction proceeds *via* **VII/29** as
shown in Scheme VII/7. Catalytic hydrogenation and desulfonylation then give
(±)-muscone (**VII/25**) in high yield [15].

Ring enlargement reactions also take place in 2-oxocycloalkane-1-carbonitri-
les substituted in 1-position by an ω-alkylester or ketone [17]. The introduction
of cyano groups into the α-position of cycloalkanones can be carried out in
CH₂Cl₂ with ClSO₂NCO in dimethylformamide [18]. Two and three carbon
atom ring expansion reactions are possible by this method. In most cases the
yields are low, which is in contrast with the results of the lactonisation (compare

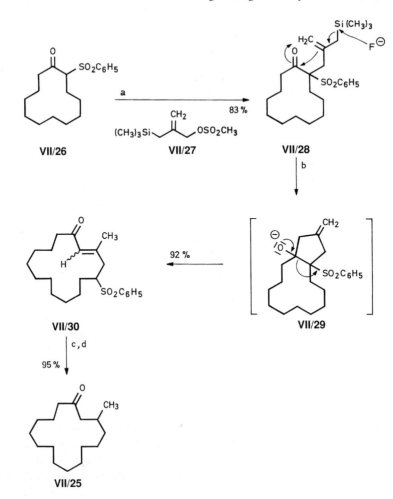

Scheme VII/7. (±)-Muscone (**VII/25**) synthesis mediated by phenylsulfone [15].

a) NaH, NaI, 1,2-dimethoxyethane   b) Bu₄NF, THF
c) H₂, Pd/BaSO₄   d) Na(Hg), Na₂HPO₄.

Chapter VII.3). The conversion of **VII/31** to **VII/32** (Scheme VII/8) however, goes in very good yield. Compound **VII/33** is formed by transesterification [17]. One, two, and three carbon atom expansions, following the same reaction principle, have been published quite recently [19] [19a]. By this method, ethyl 1-methoxycarbonylmethyl-2-oxocyclohexancarboxylate, for example, can be transformed to ethyl 2-methoxycarbonyl-3-oxocycloheptancarboxylate in the presence of 1.2 equivalents of potassium tertiary butoxide in dimethylsulfoxide (41 % yield).

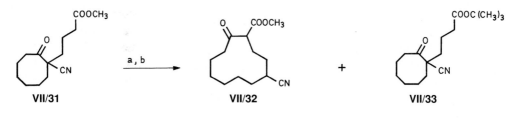

Scheme VII/8. a) KOtBu, THF   b) HOAc.

Using these methods it should be possible to carry out the following interesting reaction sequence: Suppose one uses a nucleophile which can also act as a leaving group. Two or three different types of Michael acceptors (one should be cyclic) may be combined in one pot. After workup, the ring enlargement should be completed. Experiments on this have not yet been completed [20]. Such a multicomponent one-pot annulation [20] [21] may start with an α,β-unsaturated cycloalkanone, e.g. **VII/34**, (Scheme VII/9). In a series of reactions only Michael additions take place. The whole sequence is named "MIMIMIRC" (= Michael-Michael-Michael-Ring Closure) [20]. First, the reaction of **VII/34**

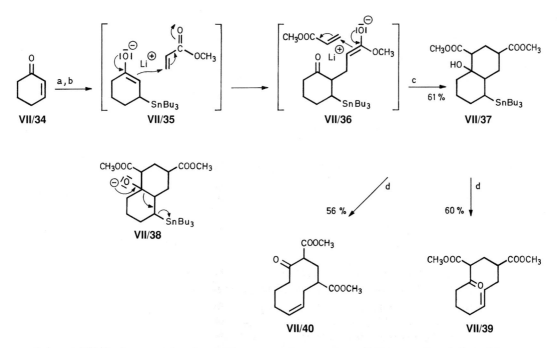

Scheme VII/9. An example of a multicomponent one-pot annulation reaction followed by oxidative cleavage [20].

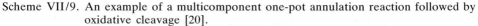

a) LiSnBu₃, THF   b) 2 CH₂=CH-COOCH₃   c) NH₄Cl, H₂O   d) Pb(OAc)₄.

with lithium tributyl stannate leads to the lithium enolate, **VII/35**. This compound reacts with methyl acrylate to form again a lithium enolate, the intermediate **VII/36**, which reacts with a further Michael acceptor, added later on (Scheme VII/9). In Scheme VII/9 methyl acrylate is used twice afterwards. The newly formed carbanion attacks the old cyclohexene carbonyl group. It had been planned that this tertiary alkoholate would start a fragmentation type elimination, in which the original nucleophile (SnBu$_3^\ominus$) should act as a leaving group, **VII/38**. However this could not be realized without a further chemical transformation of the "leaving group". This was realized by lead tetraacetate oxidation of the two diastereoisomers, **VII/37**, separated from the reaction mixture. The reaction sequence resembles a boomerang.

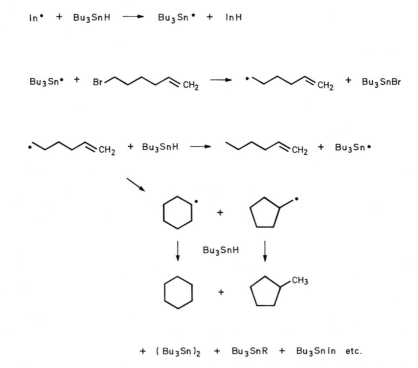

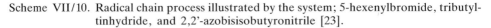

Scheme VII/10. Radical chain process illustrated by the system; 5-hexenylbromide, tributyl-tinhydride, and 2,2'-azobisisobutyronitrile [23].

In· = initiator radical.

The use of organic tin compounds leads to another aspect of the same topic. An effective alternative to the ring enlargement following ionic reaction mechanisms are those expansions which are radical chain processes. The stannylation and destannylation of organic molecules [22] as well as the knowledges of radical chain processes[4] have been used in some cases.

In a free radical ring expansion reaction discovered quite recently [24] [25], a methyl or ethyl cycloalkanone-2-carboxylate was first alkylated with methylenedibromide [24] or bromomethylphenylselenide [25], Scheme VII/11.

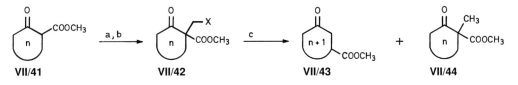

VII/41   **VII/42**   **VII/43**   **VII/44**

Scheme VII/11. Ring expansion of bromo- or selenomethyl β-ketoesters [24] [25].

X = Br, SeC$_6$H$_5$
a) NaH, THF, hexamethylphosphoramide   b) CH$_2$Br$_2$ or Br-CH$_2$SeC$_6$H$_5$
c) Bu$_3$SnH, benzene, 2,2'-azobisisobutyronitrile, reflux.

Treatment of the alkylation product, **VII/42**, with tributyltinhydride and a catalytic amount of 2,2'-azobisisobutyronitrile in refluxing benzene gives the product **VII/43**, expanded by one carbon atom[5]. In the case of methyl cyclopentanone-2-carboxylate, the yield of the bromomethyl adduct **VII/42** is 67 % and the ring enlargement proceeds in 75 %. The expected reduction product, **VII/44**, was not isolated [24]. The proposed mechanism is given in Scheme VII/12 [24] [26]. The ring expansion reaction probably occurs when the primary radical attacks the ketone carbonyl carbon atom. The resulting alkoxy radical forces the internal cyclopropane ring, **VII/47**, bond to cleave. The ester group plays several important roles in this rearrangement: By double activation the synthesis of the starting materials is facilitated. It also appears to activate the ketone towards attack by the methylene radical. At a later stage the ester stabilizes the radical in its α-position through conjugation and so provides the driving force for cyclopropane ring cleavage.

Further results using homologous bromides and iodides, treated under the same conditions, are given in Scheme VII/13. Ring enlargement by two carbon atoms was not achieved [31]. The only product isolated from the reaction mixture was the reduced ethyl 2-ethyl-2-cyclohexanonecarboxylate (**VII/52**).

---

4) Although radical chain processes occur spontaneously at moderate temperatures, it is usually desirable to faciliate the chain propagation by addition of an initiator. 2,2'- Azobisisobutyronitrile (AIBN = 2,2'-dimethyl-2,2'-azobis[propanenitrile]) is an ideal initiator, its decomposition rate is solvent independent. Such a reaction is described for a prototype 5-hexenylbromide with tributyltin hydride initiated by "In" [23] as an illustration in Scheme VII/10.
5) An open chain (thioester) analogue of this migration is also known [27] [28] [29] [30].

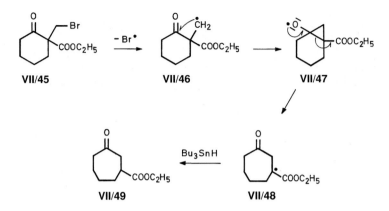

Scheme VII/12. Proposed mechanism of the ring expansion by a radical reaction [26].

In contrast, three- and four-carbon atom ring expansions proved more success-ful. Similar results were observed in case of cyclopentanones and cyclohepta-nones [31], and in various heterocyclic systems (N, O, S) [32].

Analogous ring enlargement reactions, including a radical promoted C_1 incorporation, are known in steroid chemistry: Irradiation of the 11β-nitrite of 4-androstene-11β-ol-3,17-dione (**VII/59**) (Barton reaction) in toluene gave 18-nor-D-homo-4,13(17a)-androstadiene-11β-ol-3,17-dione (**VII/63**) [33]. For C_1 radical rearrangements mediated by cobalamin, see [34] [35] [36] [37].

A related reaction was observed when the cyclic keto ester, **VII/64**, (Scheme VII/14) was boiled with tributyltinhydride in benzene (or toluene). The result-ing product was the benzocyclooctanone, **VII/68** [38]. As expected, this com-pound is generated *via* formation and β-fission of the alkoxy radical, **VII/66**, but the yield is low. The major product is the direct reduction product, **VII/67**, derived from the unrearranged radical **VII/65**. If the reaction is carried out with deuteriostannane, **VII/67** shows the presence of deuterium on both rings, the aryl as well as cycloalkyl. Therefore it can be concluded that not only **VII/65** but also **VII/69** is an intermediate. Homologues of the radical, **VII/69**, are supposed to be intermediates for the rearrangement products (ring-contracted and ring-expanded as well as reduction products) which can be observed in homologues of **VII/64** [25] [39].

Formation and β-fission of bicyclic tertiary alkoxyl radicals from the corre-sponding alcohols are well known [38] [40]. The treatment of 5α-cholestane-3β,5-diol-3-acetate, **VII/70**, and the 5β-alcohol, **VII/71**, respectively (Scheme VII/15), with one molar equivalent of lead tetraacetate in the presence of anhydrous calcium carbonate gives radical fragmentation reactions. The pro-ducts are the two (E)- and (Z)-3β-acetoxy-5,10-seco-1(10)-cholesten-5-ones (**VII/72** + **VII/73**) [40]. The ratio of **VII/73**:**VII/72** is 63:10 [41] [42] [43].

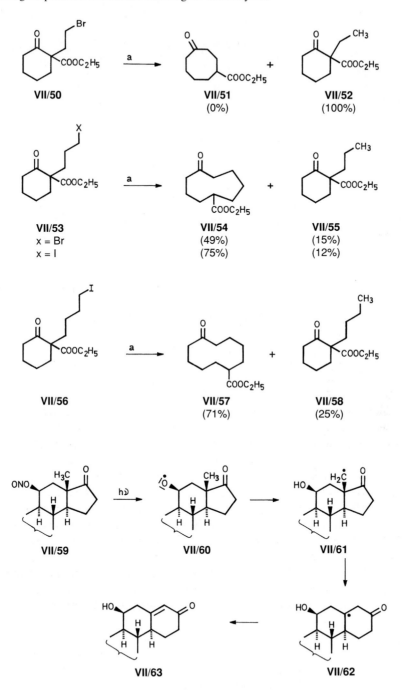

Scheme VII/13. Examples of ring expansions by radical chain processes [26].

a) Bu₃SnH, 2,2'-azobisisobutyronitrile.

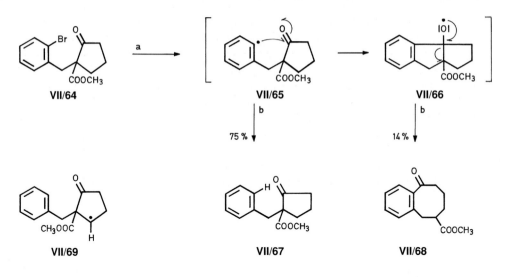

Scheme VII/14. Ring expansion by β-fission of an alkoxyl radical in a bicyclic system [25] [39].

a) Bu$_3$SnH, benzene, 2 h reflux   b) Bu$_3$SnH.

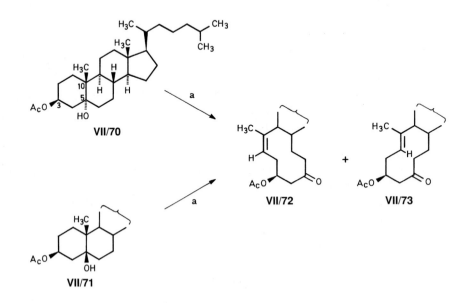

Scheme VII/15. Preparation of stereoisomeric 5,10-seco-cholestene derivatives [40].

a) Pb(OAc)$_4$, CaCO$_3$, benzene, reflux.

An alternative ten-membered ring formation is obtained by irradiation of **VII/70** in the presence of mercury(II)oxide and iodine in CCl$_4$ solution.

There exists also a synthesis of cyclopentadecanone (**VII/81**) and (±)-muscone, based on a three-carbon annulation of cyclic ketones followed by the regioselective radical cleavage of the zero bridge of the so formed bicyclic system [44]. The synthesis of cyclopentadecanone is summarized in Scheme VII/16. The cyclization of **VII/78** to the bicyclic alcohol **VII/79** proceeds best (94 % yield) with samarium diiodide in the presence of hexamethylphosphoric acid triamide and tetrahydrofuran [45]. The oxidative cleavage of **VII/79** to the ring expanded product **VII/80**, was performed by treatment with mercury(II)-oxide and iodine in benzene, followed by irradiation with a 100 Watt high pressure mercury arc. Tributyltinhydride made the de-iodination possible.

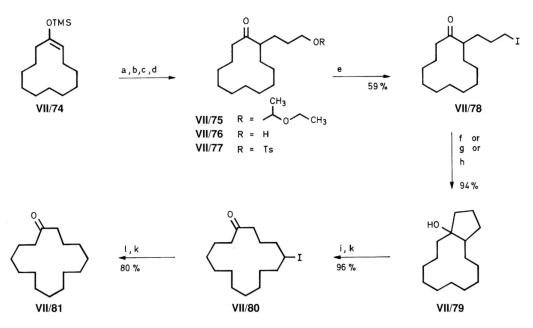

Scheme VII/16. Synthesis of cyclopentadecanone (**VII/81**) by β-fission of the bicyclic zero bridge [44].

a) CH$_3$Li, 1,2-dimethoxyethane   b) I(CH$_2$)$_3$OCH(CH$_3$)OC$_2$H$_5$ c) H$_3$O$^\oplus$
d) TsCl, pyridine   e) NaI, acetone
f) SmI$_2$, THF, hexamethylphosphoramide   g) BuLi, THF
h) Mg, HgCl$_2$, THF   i) HgO, I$_2$, benzene   k) hν
l) Bu$_3$SnH, 2,2'-azabisisobutyronitrile, benzene.

Mechanistic studies on the *cis*- and *trans*-isomer of 9-decalinoxyl radicals (Scheme VII/17), generated from a variety of reagents (*e.g.* hypobromite, nitrite), indicate a delicate balance of kinetic and thermodynamic factors influencing the direction of ring opening. Each of the isomers of the 9-decalinoxy radicals undergo fast, but reversible, 9,10-bond fissions though the 1,9-bond fission is slower than the 9,10-bond fission (tertiary radicals are preferred to secondary ones); Scheme VII/17 [46]. For related reactions, see ref. [47] [48].

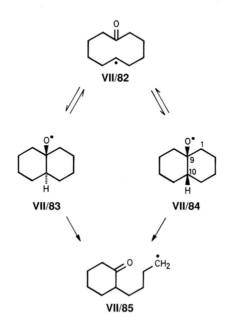

Scheme VII/17. *Cis*- and *trans*-isomers of the 9-decalinoxyl radicals [46].

Finally, a ring enlargement system will be discussed, which has been discovered quite recently [49] [50]. In some respects, it represents a combination of two methods, mentioned above [20] [26] (compare Schemes VII/9 and VII/13). 2-Cyclohexanone was substituted in the 3-position by a tributyl-tin residue, and in the 2-position by an ω-functionalized side chain, see Scheme VII/18. Using this method, the side chain could be introduced regioselectively (*trans* to the SnBu$_3$ group). Starting with the acetal, **VII/89**, and reversing the order of introduction for the groups gave a product, **VII/91**, with the reversed configuration at the two centers.

Treatment of *cis*- and *trans*-isomers with lithium tributyltinhydride in benzene solution with 2,2'-azabisisobutyronitrile as initiator gave ring enlargement products, with only small amounts of direct reduction products. Because of the different geometry of the intermediates, the reaction leads, as shown in Scheme VII/19, to different products. The *cis*-substrate **VII/91** (X = SeC$_6$H$_5$, Y = CH$_3$: *cis*-2-methyl-2-(4-phenylselenobutyl)-3-tributylstannylcyclohexanone, Scheme VII/18) forms the *cis*-alkene (**VII/94**, (*Z*)-6-methylcyclodec-5-enone) in 89 % yield. Compound **VII/94** was contaminated with approximately 10 % of

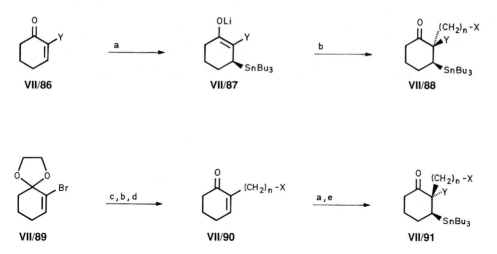

Scheme VII/18. Synthesis of *cis*- and *trans*-2-(phenylselenoalkyl)-3-(tributylstannyl)cyclo-hexanones [50].

Y = CH₃ or R
a) Bu₃SnLi, THF   b) I-(CH₂)ₙ-X   c) BuLi
d) (COOH)₂, H₂O e) Y⊕.

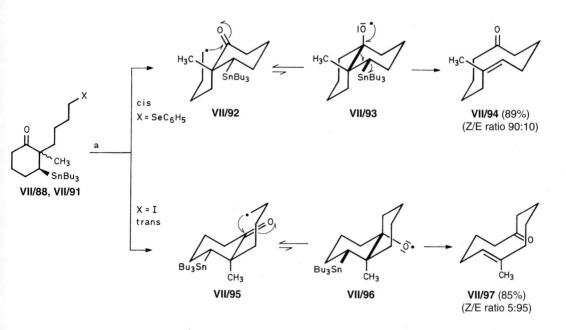

Scheme VII/19. A free radical mediated ring expansion of *cis* and *trans* α-alkylated β-stanny-lated cyclohexanones [49] [50].

a) Bu₃SnLi, 2,2'-azabisisobutyronitrile, benzene, reflux.

the (*E*)-isomer. The *trans*-substrate, **VII/88** (X = I, Y = CH$_3$), produces the (*E*)-alkene, **VII/97**, (85 % yield) with an (*E*)/(*Z*) ratio greater than 95:5. The reaction mechanism shown in Scheme VII/19 agrees with these observations: Elimination of the trialkylstannyl radical proceeds *via* a concerted coplanar *anti*-elimination mechanism and is an alternative to the fragmentation reactions of similar systems (Chapter VIII) [50].

Free radical reactions have been used in organic synthesis not only for ring expansion, but also for formation of macrocyclic ketones [48] [51].

VII.2. Ring Enlargement by Side Chain Incorporation with Lactam Formation

As explained in the previous sections, a carbanion or a carbon radical situated in the side chain of a cycloalkanone derivative can react with the carbonyl group to give a expanded ring system *via* a bicyclic intermediate. If the nucleophilic side chain carbon atom is replaced by a nitrogen atom, the expected product will be a lactam. Reactions are known in which the additional ring in the bicyclic intermediate is six- [52] [53] and seven-membered [54], respectively. Some of these results, observed in cyclododecanone derivatives, are presented in Scheme VII/20. The amine is formed from the reaction of the aldehyde group in 3-(1-nitro-2-oxocyclododecyl)propanal (**VII/98**) with ammonia or an amine followed by reduction either with NaBH$_3$CN[6] in tetrahydrofurane or with NaBH$_4$ in ethanol. The enlargement *via* a six-membered ring (**VII/98 → VII/99**) can be catalyzed by NaHCO$_3$, while the one *via* a seven-membered ring (**VII/102 → VII/103**) needs more drastic conditions, Scheme VII/20.

Compared to t.l.c. results the effective yields in the two step process seem to be much better than if the products are isolated and purified. There are indications that during chromatography, some of the nitro lactams are destroyed. In competition with this reductive amination/ring enlargement reaction is the enamine expansion route, discussed at the beginning of this Chapter (Scheme VII/6). Therefore the imine in this sequence must be reduced immediately after its formation.

The synthesis of desoxoinandenine (**VII/108**) represents an application of this lactam formation process [55] [56], Scheme VII/21. Desoxoinandenine is a reduction product of the natural spermidine alkaloid inandeninone from *Oncinotis inandensis* Wood et Evans [57].

---

6) If this reductive amination is carried out in NaCNBH$_3$/C$_2$H$_5$OH or CH$_3$OH, alkoxy derivatives of the corresponding lactams are formed *e.g.* 15-methoxy-12-nitro-15-pentadecanelactam from **VII/98** and NH$_4$OAc, CH$_3$OH, NaBH$_3$CN in 46 % [53].

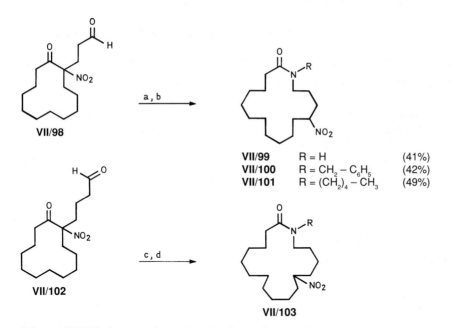

VII/99   R = H   (41%)
VII/100   R = CH₂ – C₆H₅   (42%)
VII/101   R = (CH₂)₄ – CH₃   (49%)

Scheme VII/20. Lactam formation by ring enlargement via Schiff bases.

VII/103: R=(CH₂)₃N(SO₂C₆H₅)(CH₂)₄NH(SO₂C₆H₅) 40 %
a) H₂NR, THF, NaBH₃CN   b) NaHCO₃, H₂O, CH₃OH, 20°
c) H₂NR, C₂H₅OH, NaBH₄   d) KH, [18]crown-6, 1,2-dimethoxyethane.

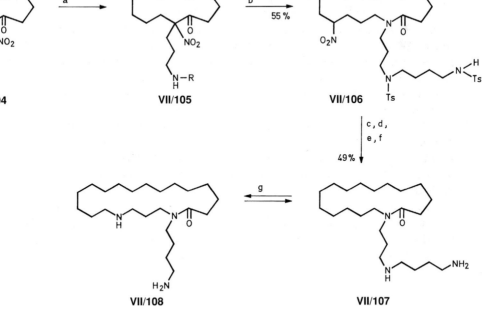

Scheme VII/21. Synthesis of desoxoinandenine (**VII/108**) [55] [56].

R = (CH₂)₃N(Ts)(CH₂)₄NHTs
a) H₂NR, NaBH₃CN   b) NaHCO₃, H₂O   c) NaOCH₃, TiCl₃, NaOAc, H₂O
d) BF₃, (CH₂-SH)₂   e) Raney-Ni   f) electrolysis   g) TsOH, xylene.

An imide formation procedure is given in literature which includes a new ring enlargement concept [58], Scheme VII/22.

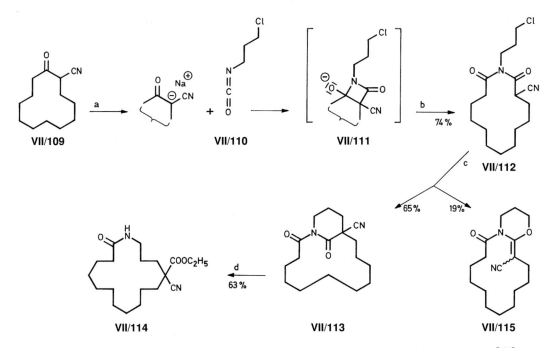

Scheme VII/22. Ring enlargement of active methylene compounds with isocyanates [58].

a) NaH, THF   b) H₂O   c) K₂CO₃, dimethylsulfoxide   d) C₂H₅OH.

Treatment of the sodium salt of 2-cyanocyclododecanone (**VII/109**) with ω-haloisocyanates yields the 14-membered imide **VII/112**. Two cyclization products of **VII/112** were obtained in the presence of potassium carbonate as base. In the C-alkylated bicycle, **VII/113**, the central bridge bond is solvolyzed to form the 16-membered amide **VII/114**. The O-alkylated bicycle, **VII/115**, which is the minor product, was not investigated further. When 2-chloroethyl isocyanate was used as reagent, the analogue of **VII/115** was formed directly. Because of its multistepcharacter the reaction resembles MIMIRC (Scheme VII/9) [59] [60].

VII.3. Lactone Formation by Side Chain Incorporation

Various strategies has been used to synthesize macrocyclic lactones by ring enlargement. The reason for this development lies in the large number of different naturally occurring macrocyclic lactones. Many of these are of considerable clinical importance [61] [62] [63] [64] [65].

One of the reactions extensively investigated is the heterolytic cleavage of the C(1),C(2) bond of cycloalkanones substituted in 2-position by an electron withdrawing group, Y, Scheme VII/23.

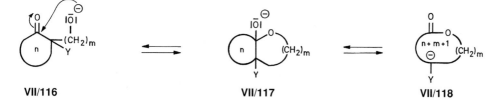

VII/116   **VII/117**   **VII/118**

Scheme VII/23. Lactone formation by heterocyclic ring enlargement.

Y = CN [98], C=O [101], NO$_2$ [95], SO$_2$C$_6$H$_5$ [16]
n = ring size: 6 – 16
m = number of carbon atoms in the side chain.

Attack of the side chain alkoxide at the carbonyl group in **VII/116** leads to the bicyclic hemiacetal anion **VII/117**, which openes under the influence of the substituent Y, to give the lactone anion **VII/118**. Depending on several factors (ring size, nature of the electron withdrawing group Y, nature of the cation *etc.*) the three different anions can be in a thermodynamic equilibrium [1]. Depending on this equilibrium, the configuration at the bridge head in the bicyclic intermediate **VII/117** can change during the reaction. As mentioned at the beginning of this chapter, individual intermediates have been isolated, and their configurations have been determined [3] [4]. However, because of a possible equilibration between the isolated and an unknown intermediate, it is not advisable to predict the stereochemistry of the conversion using the structures of the bicycles of type **VII/117**.

2-Nitrocycloalkanones have been successfully used for the preparation of many natural products.

The ten-membered lactone (R)-(-)-phoracantholide I [(-)-**VII/126**] [66] was first isolated from the metasternal gland of the eucarypt longicorn *Phoracantha synonyma* Newman [67]. It is part of the beetle's defensive secretion. Several times in the past this compound has been synthesized several times using ring forming [68] [69] [70] [71] [72] [73], as well as ring enlargement reactions [3] [74] [75] [76] [77] [78] [79] [80] [81] [82].

The starting material in the synthesis given in Scheme VII/24 is 3-(1-nitro-2-oxocyclohexyl)propanal (**VII/119**) prepared from 2-nitrocyclohexanone and acrylaldehyde in the presence of triphenylphosphine [3] [83]. The chemoselective methylation of the aldehyde group is possible using dimethyltitanium-diisopropoxide. The only isolable product is the hemiacetal **VII/124** (structure by X-ray crystallography). No other diastereoisomer of **VII/124** could be detected. The ring enlargement of the bicyclic **VII/124** is carried out using catalytic amounts of tetrabutylammonium fluoride in tetrahydrofuran to get the ten-membered compound **VII/125**, a mixture of diastereoisomers, in nearly quantitative yield. In a Nef type reaction the secondary nitro group is transformed to a ketone and then reduced to (±)-phoracantholide I ((±)-**VII/125**) [3]. The reduction of the secondary nitro group can also be achieved with tributyltin-hydride [77].

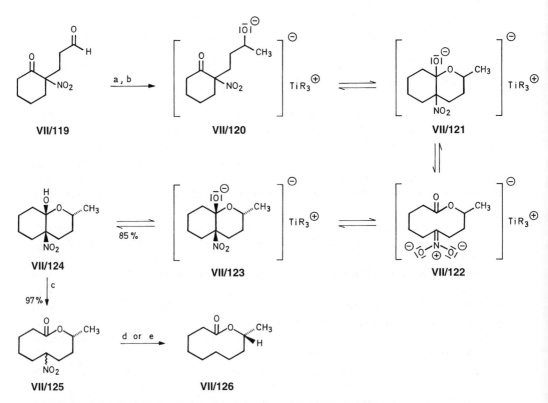

Scheme VII/24. Synthesis of (±)-phoracantholide I ((±)-**VII/126**) [3] [8].

a) (CH₃)₂Ti(iOPr)₂, (C₂H₅)₂O   b) KF, H₂O   c) Bu₄NF, THF
d) LiOCH₃, CH₃OH, KMnO₄, Na₂B₄O₇, (HSCH₂)₂, BF₃, Raney-Ni, CH₃OH, **VII/125** → **VII/126**: 69 % [3]
e) Bu₃SnH, 2,2'-azabisisobutyronitrile, 110°, toluene. **VII/125** → **VII/126**: 48 % [77]

Of interest from the mechanistic point of view is the formation of only one diastereoisomer in the methylation step **VII/119** → **VII/124**. Two possible explanations are discussed in the literature [3]. First, a stereoselective methylation of the aldehyde group takes place under the influence of the nitro group leading to the correct stereochemistry in **VII/124**. The second possibility involves the titanium reagent. An equilibrium can exist between the diastereoisomeric mixture **VII/121** and the pure **VII/123** *via* the isomer **VII/122**. By quenching the equilibrium mixture, only the thermodynamically most stable isomer would be obtained [3]. A differentiation of the two mechanisms seems possible using chiral reaction conditions. Treatment of the chiral (-)-**VII/119** (50 % ee), prepared by an asymmetric Michael addition of acrylaldehyde and 2-nitrocyclohexanone in the presence of cinchonine [84], with achiral dimethyltitaniumdiisopropoxide yields only achiral methylation products. This experiment shows that no stereoselective methylation takes place. The second consideration, then seems to be more likely (Scheme VII/24)[7].

In the course of the synthesis of (-)-15-hexadecanolide ((-)-**VII/129**) a 1,4-interaction of the nitro and the ketone carbonyl group was observed [3] [86], Schema VII/25. The desired compound was prepared from 4-(1-nitro-2-oxo-cyclododecyl)butan-2-one (**VII/127**) by reduction of the ketone to an alcohol followed by ring enlargement to the 16-membered nitro-lactone. From this the nitro group had to be removed reductively. 15-Hexadecanolide contains one center of chirality which is established by the reduction of **VII/127** to **VII/128**. This reduction was done with the organoboron complexes, (S)-Alpine-Hydride, and (R)-Alpine-Hydride [87], as well as with sodium borohydride. It proved to be regiospecific in all cases; in tetrahydrofurane at −78° for 2 h or, with sodium borohydride, in methanol at 0° for 4 h. Characterization of the direct reduction product **VII/128** was not possible, because **VII/128** was converted partly or completely into the ring enlarged compounds 12-nitro- and 12-oxo-15-hexadecanolide during the reduction and the workup. In order to compare the results, all intermediates have had to be transformed into the lactone **VII/129** [86]. (R)-Alpine-Hydride is known to reduce 2-butanone to (-)-(R)-2-butanol [87]. If (±)-**VII/127** is the starting material, the (S)- and (R)-Alpine-Hydride give the corresponding (S)- and (R)-configurated lactones (+)-**VII/129** and (-)-**VII/129**, respectively. Reduction of (+)-**VII/127** (optical purity better than 95 %) with sodium borohydride gave (+)-**VII/129** with 14 % ee. Reduction of (+)-**VII/127** with (S)-Alpine-Hydride gave (+)-(S)-**VII/129** (74 % ee). An unexpected result was the formation of (S)-**VII/129** (45 % ee), prepared from (+)-**VII/127** with (R)-Alpine-Hydride. The nitro group seems to form a complex with the organoboron compound. The complex (S)-Alpine-Hydride/(+)-**VII/127** is preferred to the alternative (R)-Alpine-Hydride/(+)-**VII/127**. Molecular models indicate that the reduction of (+)-**VII/127** with borohydrides yields, independent of

---

7) Optical active phoracantholide I (**VII/126**) was observed when (±)-2-nitro-2-(3'-oxobutan-4-yl)cyclohexanone was reduced with (S)- or (R)-Alpine-Hydride [85].

their configuration predominantly (+)-(15S)-**VII/129**. From these results the conclusion can be drawn, that the complex between the nitro group and the boron atom is more important than the one between the boron atom and the carbonyl group, at least with respect to the newly formed chiral center [86], compare [88] [89].

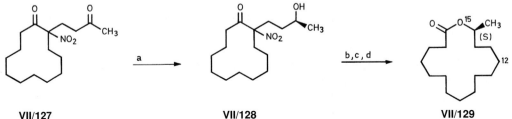

VII/127   **VII/128**   **VII/129**

Scheme VII/25. Synthesis of (+)-15-hexadecanolide ((+)-**VII/129**) [86].

a) Reduction   b) Bu$_4$NF, THF   c) TiCl$_3$, NaOAc
d) Zn, HCl.

	Entry	Reduction conditions	yield [%]	**VII/129** ee [%]	configuration at C(15)
(±)-**VII/127**	1	(S)-Alpine-Hydride	82	15	(S)
	2	(R)-Alpine-Hydride	72	24	(R)
(+)-**VII/127**	3	NaBH$_4$	84	14	(S)
	4	(S)-Alpine-Hydride	76	74	(S)
	5	(R)-Alpine-Hydride	40	45	(S)

The natural products [90], dihydrorecifeiolide (= 11-dodecanolide) [3] [91], 12-methyl-13-tridecanolide [92], and 12-tridecanolide [93] have been synthesized in high overall yields using this ring enlargement reaction as a key step.

So far, no systematic investigation of the length of the side chain in **VII/116** (Scheme VII/23) has been made. There have been, however, enquieries into lactonization reactions, in which the side chain contained two [94], three [4] [83] [95] [96], and four [93] carbon atoms. In these cases, the intermediate hemiacetal ring is a five-, six-, or seven-membered ring. The yields of the lactones, generated *via* five- and six-membered hemiacetals, are usually high (90% and more). If a lactonization reaction involves a seven-membered hemiacetal, the yields are distinctly lower.

This statement is only true if the intermediate seven-membered ring contains no double bond. Using the phenylsulfonyl residue as an electronegative group

(Scheme VII/23) the allylic alcohol **VII/130**, prepared from the corresponding 2-phenylsulfonyl-cycloalkanone and (Z)-4-chloro-2-buten-1-yl acetate, is transformed to the (Z)-lactone **VII/131** in good yields (from six-membered **VII/130** 76%, eight 79%, twelve 89%) [16], Scheme VII/26.

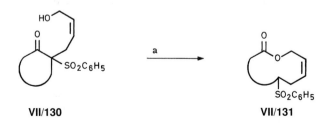

Scheme VII/26. Ring enlargement reaction using a (Z)-configured side chain [16].

a) NaH, C$_6$H$_6$, Δ.

An interesting approach to macrocyclic benzolactones was discovered by treating 2-nitrocycloalkanones with 1,4-benzoquinone (**VII/133**) in the presence of catalytic amounts of 1,8-diazabicyclo[5.4.0]undec-7-ene (DBU). The transformation involves a Michael reaction, aromatization, and ring enlargement *via*

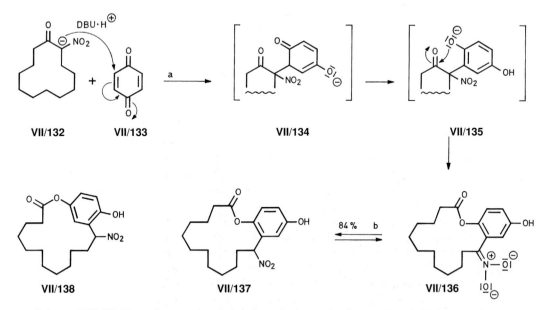

Scheme VII/27. Benzolactone formation from 2-nitrocycloalkanones and 1,4-benzoquinone [94].

a) 1,8-Diazabicyclo[5.4.0]undec-7-ene, THF, 20°   b) H$_2$O.

a five-membered intermediate. In Scheme VII/27, the formation of 17-hydro-xy-14-nitro-2-oxabicyclo[13.4.0]nonadeca-1(15),16,18-trien-3-one (**VII/137**) is shown [94]. The alternate 16-membered **VII/138** was not observed.

The electron withdrawing capacity of the nitrile group can be used to a series of syntheses of ring enlarged lactones in high yields [97] [98] [99]. For example, the preparation of 12-methyl-15-pentadecanolide (**VII/143**) from 1-(2-formyl-ethyl)-2-oxocyclododecane-1-carbonitrile (**VII/139**) is summarized in Scheme VII/28 [99].

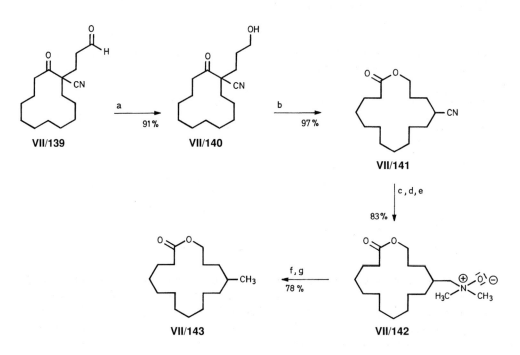

Scheme VII/28. Synthesis of enlarged lactone by use of carbonitrile [99].

a) NaBH$_4$, CH$_3$OH b) Bu$_4$NF, THF c) H$_2$, PtO$_2$, C$_2$H$_5$OH
d) CH$_2$O, H$_2$O, dioxane, NaH$_2$PO$_3$
e) *meta*-chloroperoxybenzoic acid, CHCl$_3$
f) 150°, 0,005 Torr dest. g) H$_2$, Pd-C, C$_2$H$_5$OH.

Replacement of the carbonitrile by the ethoxycarbonyl group leads to compounds with two electrophilic centers (ketone and ester carbonyl) of similar order. With these compounds, base catalyzed ring enlargement was not observed. The main products were explained by attack of the nucleophile at the ethoxycarbonyl group [79].

The cleavage of cyclic 1,3-dicarbonyl compounds has been used extensively for the preparation of long-chain carboxylic acids and esters [100]. The intra-molecular version of this reaction is a ring enlargement. For this purpose an intramolecular nucleophile, a hydroxyl group, placed in the side chain, which is located in the 2-position between the two carbonyl groups, reacts by a retro Dieckmann reaction to form ketone-lactones. A number of such lactone syntheses have been reported, see Scheme VII/29.

It is remarkable that 1,3-cyclohexanediones with side chains containing seven, ten, and even twenty-one members are ring enlarged under basic anhydrous reaction conditions. The yields of the resulting 13-, 16-, and 27-membered ketone-lactones, are given in the following Table [104]. As expected, dimerized dilactones are side reaction products[8].

Table. Lactone formation by incorporation of long side chains in cyclic 1,3-dicarbonyl compounds [104].

Number of CH_2 groups in the side-chain (n) of **VII/150**	hemiacetal ring size	ring size of the resulting ketone-lactone	Yield [%]
4	7	11	49
6	9	13 (**VII/151**)	28
9	12	16 (**VII/152**)	56
20	23	27 (**VII/153**)	11

The introduction of olefinic subunits into ketone-lactones by ring enlargement is only possible, when the configuration of the double bond is (Z), as shown in Scheme VII/29, (conversions of **VII/144** → **VII/145**). Direct application with an acetylenic analogue is not possible geometrically, because of the linear nature of the butynyl subunit. However, complexation of the C,C triple bond with a metal can change the geometry dramatically [105] and make the reaction possible. A reaction with $Co_2(CO)_8$ [106], for example, results in a geometric change to a system with approx. 140° bond angles around the alkyne carbon atoms. Thus, a complexed triple bond in a side chain of a 1,3-diketone, should yield a system capable of ring enlargement. Such molecules do react as desired. The complexed triple bond behaves like a (Z) olefinic bond. The $Co_2(CO)_6$ complex **VII/155**, *e.g.*, prepared as outlined in Scheme VII/30, undergoes ring enlargement to an eleven-membered complex, **VII/156**, in the presence of 1 equivalent of sodium hydride in 1,3-dimethoxyethane at room temperature. Even in the absence of base, a pure sample of **VII/155** stored at 0° for two weeks, underwent a spontaneous lactonization [107].

---

8) Further examples are given in [88].

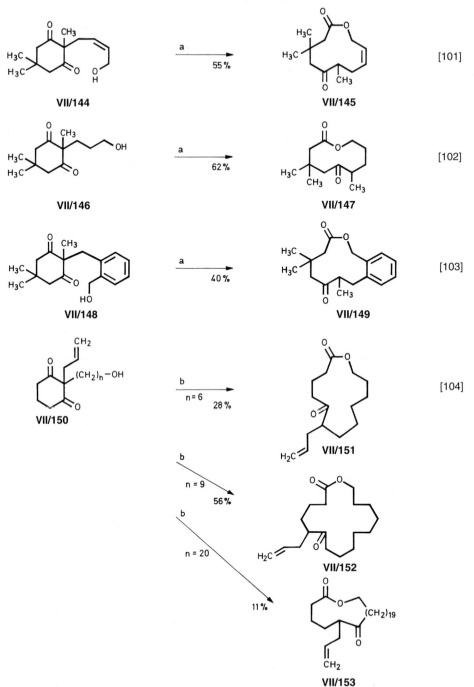

Scheme VII/29. Further examples of lactone formation by ring enlargement. The order of the substituents at C(5) and C(6) in **VII/152** and **VII/153** is reversed.

a) NaH, C$_6$H$_6$, 80°   b) NaH, THF.

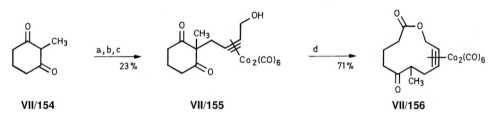

VII/154   **VII/155**   **VII/156**

Scheme VII/30. Synthesis of lactones containing triple bonds [107].

a) Br-CH$_2$-C≡C-CH$_2$-OSi(Bu)(CH$_3$)$_2$ b) 5 % HF, CH$_3$CN
c) Co$_2$(CO)$_8$ d) 1 eq. NaH, 1,2-dimethoxyethane, 20°.

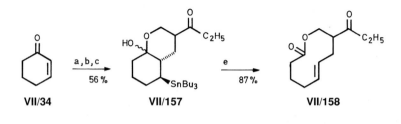

VII/34   **VII/157**   **VII/158**

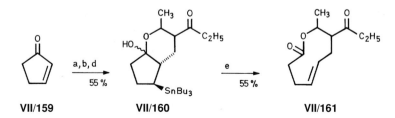

VII/159   **VII/160**   **VII/161**

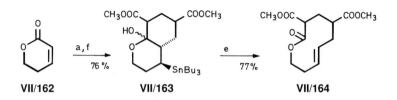

VII/162   **VII/163**   **VII/164**

Scheme VII/31. Examples of one-pot four-component annulation [74].

a) LiSnBu$_3$, THF   b) CH$_2$=CH-COC$_2$H$_5$ c) CH$_2$O   d) CH$_3$CHO
e) Pb(OAc)$_4$ f) 2 CH$_2$=CH-COOCH$_3$.

A multicomponent one-pot annulation reaction, such as the one described in Chapter VII.1, can also be used to synthesize lactones with enlarged rings. [74] [108].

The formation of the bicyclic intermediate, **VII/157**, (Scheme VII/31) was achieved by nucleophilic conjugate addition of tributyltinlithium to cyclohexenone, reaction of the intermediate ketone enolate ion with a small excess of ethyl vinyl ketone and, subsequently, with a large excess of formaldehyde. The mechanism is analogous to that presented in Scheme VII/9 of Chapter VII.1.

Lead tetraacetate oxidation gave the ten-membered lactone, **VII/158**, with an overall yield of 49 %. Using the same methodology, 2-cyclopentenone (**VII/159**) and 2-cycloheptenone were converted into the nine-membered **VII/161** and the eleven-membered lactone, respectively. The yield of the cyclopentenone conversion was lower than in the case of the corresponding cycloheptenone [74]. – It was demonstrated that this type of four atom ring enlargement reaction can be used on an α,β-unsaturated lactone to give a lactone expanded by carbon atoms to an enlarged lactone: **VII/162** → **VII/163** → **VII/164**. A large number of mono- and disubstituted medium-sized lactones with an (E)-oriented double bond can be prepared regiospecifically by variations of this reaction [74][9].

As an alternative to the lead tetraacetate oxidation, (diacetoxyiodo)benzene can be used to initiate a fragmentation reaction which leads to unsaturated medium-sized lactones [110]. The structures of the starting materials are similar to those of compounds **VII/157**, **VII/160**, and **VII/163**. The same stereochemical consequences are observed as mentioned above.

Radical initiated fragmentation reactions were used for the synthesis of ring enlarged lactones (already discussed in VII.1). Several modifications and applications of this type are reported in the literature [75] [80] [111] [112] [113] [114] [115].

_____

9) Treatment of γ-hydroxyalkyl stannanes with lead tetraacetate in refluxing benzene leads to (E)- and (Z)-keto olefines in a stereospecific manner, according to the configuration of the starting material.

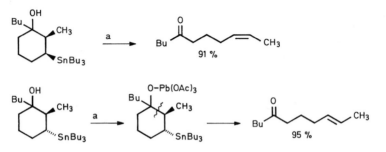

a) Pb(OAc)$_4$, benzene, 5 min, reflux.
Organostannane substituents seem to stabilize β-cations or radicals through σ-π conjugation. They undergo spontaneous elimination to form the corresponding olefins [109].

As indicated in Scheme VII/32, cyclononanone (**VII/165**) is transformed into hydroperoxide hemiacetal, **VII/167**, which is isolated as a mixture of stereo-isomers. The addition of Fe(II)SO$_4$ to a solution of **VII/167** in methanol saturat-ed with Cu(OAc)$_2$ gave (±)-recifeiolide (**VII/171**) in quantitative yield. No iso-meric olefins were detected. In the first step of the proposed mechanism, an electron from Fe^{2+} is transferred to the peroxide to form the oxy radical **VII/168**. The central C,C-bond is weakened by antiperiplanar overlap with the lone pair on the ether oxygen. Cleavage of this bond leads to the secondary carbon radical **VII/169**, which yields, by an oxidative coupling with Cu(OAc)$_2$, the alkyl copper intermediate **VII/170**. 'If we assume that the alkyl copper intermediate, **VII/170**, exists (a) as a (Z)-ester, stabilized by n (ether O) → σ^*(C=O) overlap (anomeric effect), and (b) is internally coordinated by the ester to form a pseudo-six-membered ring, then only one of the four β-hydro-gens is available for a syn-β-elimination.' [111]. This reaction principle has been used in other macrolide syntheses, too [112] [113].

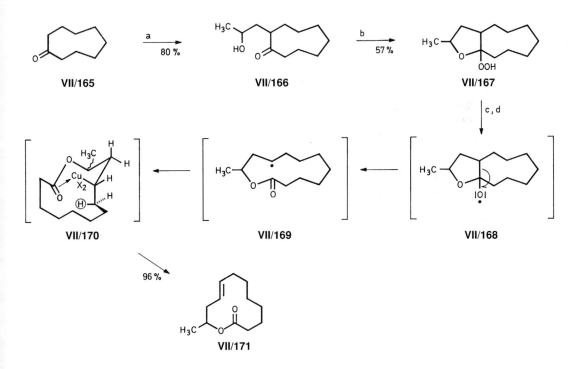

Scheme VII/32. Regio- and stereoselective ring enlargement in the synthesis of (±)-recife-iolide (**VII/171**) [111].

    a) Li enolate of **VII/165** + propylene oxide, −78°, Al(CH$_3$)$_3$
    b) H$_2$O$_2$, AcOH   c) Fe(II)SO$_4$, CH$_3$OH   d) Cu(OAc)$_2$.

In another lactonisation reaction, a cyclic hemiacetal or its opened ketone-alcohol equivalent (see Scheme VII/33, structures **VII/172** and **VII/173**), is transformed to an iodo lactone **VII/175** (12-iodopentadecan-15-olide) and the isomeric **VII/174** (2-iodo-2-(3′-hydroxypropyl)cyclododecanone) by irradiation with high pressure mercury arc in the presence of HgO-I_2 in benzene solution [114] as outlined in Scheme VII/33.

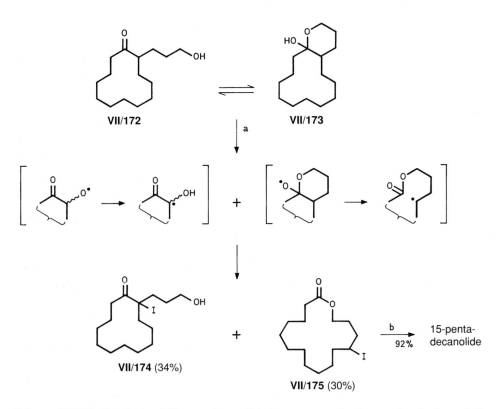

Scheme VII/33. Synthesis of 15-pentadecanolide by a consecutive intramolecular homolytic addition-β-scission of alkoxyl radicals [114].

a) HgO, benzene, I_2, pyridine, 5 h, 100 W high pressure Hg arc
b) Bu$_3$SnH, 2,2′-azabisisobutyronitrile, benzene, hν.

Both materials were isolated in approximately 30 % yield. The iodo compound **VII/175** was then reduced photochemically with tributyltinhydride to give 15-pentadecanolide [114]. Experiments with different cycloalkanones (five- to eight-membered) and different lengths of the side chain (two and three methylene groups) showed that this reaction can be used for the synthesis of several medium sized lactones [75] [80] [114]. Under similar reaction conditions, lactols can also undergo ring expansion reaction [115]. The substrates (steroidal lactols) are irradiated with visible light (100 W tungsten filament) at 40° in the presence of cyclohexane, iodine, and iodosobenzene diacetate. The reaction leads to a mixture of olefins.

VII.4. Discussion of the Auxiliary Groups

The literature contains a number of reactions, in which cycloalkanones containing additional auxiliary groups in the 2-position are used as starting materials. Some important auxiliary groups are sulfone, nitro, and cyano. The auxiliary group should be added to the cycloalkanones under mild conditions. It can then activate the 2-position for the introduction of the side chain and facilitate the heterolytic cleavage of the C(1),C(2) bond. It should also be easily removed or transformed into another functional group, after ring enlargement has taken place. The behavior of different auxiliary groups will be discussed briefly.

The 2-cyano-cycloalkanones are easy to prepare but only in moderate yields [18]. The introductions of nitro- [6] and sulfone- [5] [116] [117] groups are simpler than the cyano group, and the yields are better. Sulfone and cyano compounds are most suitable for the introduction of the side chain. Beside the Michael reaction [16] [97], and the Pd(O) catalyzed addition [15], and the reactions with alkyl halides [16] [17] proceed in good yields. In contrast to other compounds, 2-nitroketones generally do not undergo nucleophilic substitution with non-activated alkyl halides. However, Michael addition products [2], as well as products synthesized by Pd(O) catalyzed alkylation [118], are well known derivatives of 2-nitrocycloalkanones.

Subsequent transformations of functional groups in the side chain are always problematic, since the C(1),C(2) bond in the activated cycloalkanones is unstable to acid and base as well as to external nucleophiles [7] [85]. The electron acceptor properties of the sulfone and the nitro residues both promote the ring enlargement reactions. 2-Cyano-ketones seem to be less suitable for C,C-rearrangements [17], but show good results in lactonization reactions [97] [99]. Side reactions are sometimes observed, retro-Michael reactions (especially in eight-membered nitroketones), phenyl-sulfinic acid elimination in certain rearrangements of 2-phenylsulfonyl-cycloalkanones, and nitrone formation from 2-nitrocycloalkanones [12] (by attack of an internal carbanion). Side reaction of the alkoxycarbonyl group are known; e.g. formation of spiranes [119].

The removal or transformation of the auxiliary groups in the ring expansion products should be possible under mild conditions. In this respect, the sulfonyl residue has advantages because it can be reductively eliminated by Na/Hg-Na$_2$HPO$_4$ [15] [16] [120] or by electrolysis [121] in excellent yields. The direct removal of the cyano group has been reported by oxidative [122] [123] [124] [125] [126] and reductive [127] [128] methods. The oxidation of a alkylcyano group to a cyanohydrin seems to be sensitive to a number of influences not yet understood [129]. Often the nitro group is removed, stepwise by first converting it into an oxo group. Unfortunately, most methods for this so called Nef reaction [130] are too strenuous to be applied to the ring expanded products (especially lactones). However, good results have been obtained with $TiCl_3/NaOAc$ [131] or $KMnO_4$ [132] or $SiO_2/NaOCH_3$ [133]. The keto group then can be removed using well known reactions. Methods for the direct reductive removal of the nitro group have been reviewed [134]. Available procedures, mainly based on tributyltinhydride, are normally limited to the reduction of tertiary or activated secondary nitro groups.

References

[1] H. Stach, M. Hesse, Tetrahedron **44**, 1573 (1988).
[2] Y. Nakashita, M. Hesse, Helv.Chim.Acta **66**, 845 (1983).
[3] K. Kostova, M. Hesse, Helv.Chim.Acta **67**, 1713 (1984).
[4] E. Benkert, M. Hesse, Helv.Chim.Acta **70**, 2166 (1987).
[5] R Gretler, Ph. D. Thesis, University of Zürich, 1989.
[6] R. H. Fischer, H. M. Weitz, Synthesis **1980**, 261.
[7] W. Huggenberg, M. Hesse, Helv.Chim.Acta **66**, 1519 (1983).
[8] H. Stach, M. Hesse, Helv.Chim.Acta **70**, 315 (1987).
[9] M. Vavrecka, M. Hesse, Helv.Chim.Acta **72**, 847 (1989).
[10] A. Lorenzi-Riatsch, Y. Nakashita, M. Hesse, Helv.Chim.Acta **64**, 1854 (1981).
[11] Y. Nakashita, M. Hesse, Angew.Chem. **93**, 1077 (1981), Angew.Chem.Int.Ed.Engl. **20**, 1021 (1981).
[12] W. Huggenberg, M. Hesse, Tetrahedron Lett. **30**, 5119 (1989).
[13] S. Hünig, H. Hoch, Chem.Ber. **105**, 2197 (1972).
[14] S. Bienz, M. Hesse, Helv.Chim.Acta **71**, 1704 (1988).
[15] B. M. Trost, J. E. Vincent, J.Am.Chem.Soc. **102**, 5680 (1980).
[16] V. Bhat, R. C. Cookson, J.Chem.Soc., Chem.Commun. **1981**, 1123.
[17] M. Süsse, J. Hájicek, M. Hesse, Helv.Chim.Acta **68**, 1986 (1985).
[18] B. Föhlisch, R. Herter, E. Wolf, J. J. Stezowski, E. Eckle, Chem.Ber. **115**, 355 (1982).
[19] Z.-F. Xie, H. Suemune, K. Sakai, J.Chem.Soc., Chem.Commun. **1988**, 1638.
[19a] Z.-F. Xie, H. Suemune, K. Sakai, Synth.Commun. **19**, 987 (1989).
[20] G. H. Posner, E. Asirvatham, Tetrahedron Lett. **27**, 663 (1986).
[21] G. H. Posner, Chem.Rev. **86**, 831 (1986).
[22] W. C. Still, J.Am.Chem.Soc. **99**, 4836 (1977).
[23] M. Ramaiah, Tetrahedron **43**, 3541 (1987).
[24] P. Dowd, S.-C. Choi, J.Am.Chem.Soc. **109**, 3493 (1987).
[25] A. L. J. Beckwith, D. M. O'Shea, S. Gerba, S.W. Westwood, J.Chem.Soc., Chem. Commun. **1987**, 666.
[26] P. Dowd, S.-C. Choi, Tetrahedron **45**, 77 (1989).
[27] S. Wollowitz, J. Halpern, J.Am.Chem.Soc. **106**, 8319 (1984).
[28] M. Tada, K. Inoue, M. Okabe, Chem.Lett. **1986**, 703.
[29] M. Tada, K. Inoue, K. Sugawara, M. Hiratsuka, M. Okabe, Chem. Letters **1985**, 1821.
[30] S. Wollowitz, J. Halpern, J.Am.Chem.Soc. **110**, 3112 (1988).
[31] P. Dowd, S.-C. Choi, J.Am.Chem.Soc. **109**, 6548 (1987).
[32] P. Dowd, S.-C. Choi, Tetrahedron Lett. **30**, 6129 (1989).

[33] H. Reimann, A. S. Capomaggi, T. Strauss, E. P. Oliveto, D.H. R. Barton, J.Am.Chem. Soc. **83**, 4481 (1961).

[34] W. M. Best, A.P. F. Cook, J. J. Russell, D. A. Widdowson, J.Chem.Soc., Perkin Trans. I **1986**, 1139.

[35] P. Dowd, S.-C. Choi, F. Duah, C. Kaufman, Tetrahedron **44**, 2137 (1988).

[36] M. Okabe, T. Osawa, M. Tada, Tetrahedron Lett. **22**, 1899 (1981).

[37] M. Tada, K. Miura, M. Okabe, S. Seki, H. Mizukami, Chem.Lett. **1981**, 33.

[38] A. N. Abeywickrema, A. L. J. Beckwith, J.Chem.Soc., Chem.Commun. **1986**, 464.

[39] A. L. J. Beckwith, D. M. O'Shea, S.W. Westwood, J.Am.Chem.Soc. **110**, 2565 (1988).

[40] M. L. Mihailovic, L. Lorenc, M. Gašic, M. Rogic, A. Melera, M. Stefanovic, Tetrahedron **22**, 2345 (1966).

[41] M. L. Mihailovic, L. Lorenc, V. Pavlovic, J. Kalvoda, Tetrahedron **33**, 441 (1977).

[42] M. Akhtar, S. March, J.Chem.Soc. C **1966**, 937.

[43] M. Akhtar, S. Marsh, Tetrahedron Lett. **1964**, 2475.

[44] H. Suginome, S. Yamada, Tetrahedron Lett. **28**, 3963 (1987).

[45] G. A. Molander, J. B. Etter, J.Org.Chem. **51**, 1778 (1986).

[46] A. L. J. Beckwith, R. Kazlauskas, M. R. Syner-Lyons, J.Org.Chem. **48**, 4718 (1983).

[47] H. Suginome, C. F. Liu, M. Tokuda, J.Chem.Soc., Chem.Commun. **1984**, 334.

[48] T. L. Macdonald, D. E. O'Dell, J.Org.Chem. **46**, 1501 (1981).

[49] J. E. Baldwin, R. M. Adlington, J. Robertson, J.Chem.Soc., Chem.Commun. **1988**, 1404.

[50] J. E. Baldwin, R. M. Adlington, J. Robertson, Tetrahedron **45**, 909 (1989).

[51] N. A. Porter, D. R. Magnin, B.T. Wright, J.Am.Chem.Soc. **108**, 2787 (1986).

[52] R. Wälchli, M. Hesse, Helv.Chim.Acta **65**, 2299 (1982).

[53] R. Wälchli, S. Bienz, M. Hesse, Helv.Chim.Acta **68**, 484 (1985).

[54] S. Bienz, A. Guggisberg, R. Wälchli, M. Hesse, Helv.Chim.Acta **71**, 1708 (1988).

[55] R. Wälchli, A. Guggisberg, M. Hesse, Tetrahedron Lett. **25**, 2205 (1984).

[56] R. Wälchli, A. Guggisberg, M. Hesse, Helv.Chim.Acta **67**, 2178 (1984).

[57] H. J. Veith, M. Hesse, H. Schmid, Helv.Chim.Acta **53**, 1355 (1970).

[58] V. I. Ognyanov, M. Hesse, Helv.Chim.Acta **72**, 1522 (1989).

[59] R. D. Little, R. Verhé, W.T. Monte, S. Nugent, J. R. Dawson, J.Org.Chem. **47**, 362 (1982).

[60] V. I. Ognyanov, M. Hesse, Helv.Chim.Acta **73**, 272 (1990).

[61] S. Blechert, Nachr.Chem.Tech.Lab. **28**, 110 (1980).

[62] T. G. Back, Tetrahedron **33**, 3041 (1977).

[63] K. C. Nicolaou, Tetrahedron **33**, 683 (1977).

[64] S. Masamune, G. S. Bates, J.W. Corcoran, Angew.Chem. **89**, 602 (1977), Angew. Chem.Int.Ed.Engl. **16**, 585 (1977).

[65] I. Paterson, M. M. Mansuri, Tetrahedron **41**, 3569 (1985).

[66] T. Kitahara, K. Koseki, K. Mori, Agric.Biol.Chem. **47**, 389 (1983).

[67] B. P. Moore, W.V. Brown, Aust.J.Chem. **29**, 1365 (1976).

[68] Y. Naoshima, H. Hasegawa, Chem.Lett. **1987**, 2379.

[69] Y. Naoshima, H. Hasegawa, T. Nishiyama, A. Nakamura, Bull.Soc.Chem.Jpn. **62**, 608 (1989).

[70] J. Cossy, J.-P. Pete, Bull.Soc.Chim.France **1988**, 989.

[71] B. M. Trost, T. R. Verhoeven, J.Am.Chem.Soc. **101**, 1595 (1979).

[72] H. Gerlach, P. Künzler, K. Oertle, Helv.Chim.Acta **61**, 1226 (1978).

[73] T. Takahashi, S. Hashiguchi, K. Kasuga, J. Tsuji, J.Am.Chem.Soc. **100**, 7424 (1978).

[74] G. H. Posner, K. S. Webb, E. Asirvatham, S.-s. Jew, A. Degl'Innocenti, J.Am.Chem.-Soc. **110**, 4754 (1988).

[75] H. Suginome, S. Yamada, Tetrahedron Lett. **26**, 3715 (1985).

[76] J. R. Mahajan, H. C. de Araújo, Synthesis **1981**, 49.

[77] N. Ono, H. Miyake, A. Kaji, J.Org.Chem. **49**, 4997 (1984).

[78] E. Vedejs, D.W. Powell, J.Am.Chem.Soc. **104**, 2046 (1982).

[79] R. Malherbe, D. Belluš, Helv.Chim.Acta **61**, 3096 (1978).
[80] H. Suginome, S. Yamada, Tetrahedron **43**, 3371 (1987).
[81] T. Wakamatsu, K. Akasaka, Y. Ban, J.Org.Chem. **44**, 2008 (1979).
[82] T. Ohnuma, N. Hata, N. Miyachi, T. Wakamatsu, Y. Ban, Tetrahedron Lett. **27**, 219 (1986).
[83] K. Kostova, A. Lorenzi-Riatsch, Y. Nakashita, M. Hesse, Helv.Chim.Acta **65**, 249 (1982).
[84] S. Stanchev, M. Hesse, to be published.
[85] S. Stanchev, M. Hesse, Helv.Chim.Acta **73**, 460 (1990).
[86] S. Stanchev, M. Hesse, Helv.Chim.Acta **72**, 1052 (1989).
[87] S. Krishnamurthy, F. Vogel, H. C. Brown, J.Org.Chem. **42**, 2534 (1977).
[88] J. R. Mahajan, M. B. Monteiro, J.Chem.Res. (S) **1980**, 264.
[89] M. M. Midland, J. I. McLoughlin, J. Gabriel, J.Org.Chem. **54**, 159 (1989).
[90] R. Kaiser, D. Lamparsky, Helv.Chim.Acta **61**, 2671 (1978).
[91] K. Kostova, M. Hesse, Helv.Chim.Acta **66**, 741 (1983).
[92] S. Stanchev, M. Hesse, Helv.Chim.Acta **70**, 1389 (1987).
[93] H. Stach, M. Hesse, Helv.Chim.Acta **69**, 1614 (1986).
[94] H. Stach, M. Hesse, Helv.Chim.Acta **69**, 85 (1986).
[95] K. Kostova, Thesis, University of Sofia 1986.
[96] R. C. Cookson, P. S. Ray, Tetrahedron Lett. **23**, 3521 (1982).
[97] B. Milenkov, M. Hesse, Helv.Chim.Acta **70**, 308 (1987).
[98] B. Milenkov, M. Süsse, M. Hesse, Helv.Chim.Acta **68**, 2115 (1985).
[99] B. Milenkov, A. Guggisberg, M. Hesse, Helv.Chim.Acta **70**, 760 (1987).
[100] H. Stetter, in W. Foerst (Ed.), "Newer Methods of Preparative Organic Chemistry" Vol. 2, Academic Press, New York, 1963.
[101] J. R. Mahajan, Synthesis **1976**, 110.
[102] J. R. Mahajan, I. S. Resck, Synthesis **1980**, 998.
[103] J. R. Mahajan, H. de Carvalho, Synthesis **1979**, 518.
[104] P.W. Scott, I.T. Harrison, S. Bittner, J.Org.Chem. **46**, 1914 (1981).
[105] R. S. Dickson, P. J. Fraser, Adv.Organomet.Chem. **12**, 323 (1974).
[106] N. E. Schore, M. J. Knudsen, J.Org.Chem. **52**, 569 (1987).
[107] N. E. Schore, S. D. Najdi, J.Org.Chem. **52**, 5296 (1987).
[108] G. H. Posner, E. Asirvatham, K. S. Webb, S.-s. Jew, Tetrahedron Lett. **28**, 5071 (1987).
[109] K. Nakatani, S. Isoe, Tetrahedron Lett. **25**, 5335 (1984).
[110] M. Ochiai, S. Iwaki, T. Ukita, Y. Nagao, Chem.Lett. **1987**, 133.
[111] S. L. Schreiber, J.Am.Chem.Soc. **102**, 6163 (1980).
[112] S. L. Schreiber, T. Sammakia, B. Hulin, G. Schulte, J.Am.Chem.Soc. **108**, 2106 (1986).
[113] S. L. Schreiber, B. Hulin, W.-F. Liew, Tetrahedron **42**, 2945 (1986).
[114] H. Suginome, S. Yamada, Chem. Lett. **1988**, 245.
[115] R. Freire, J. J. Marrero, M. S. Rodríguez, E. Suárez, Tetrahedron Lett. **27**, 383 (1986).
[116] J. S. Meek, J. S. Fowler, J.Org.Chem. **33**, 3422 (1968).
[117] B. M. Trost, T. N. Salzmann, K. Hiroi, J.Am.Chem.Soc. **98**, 4887 (1976).
[118] V. I. Ognyanov, M. Hesse, Synthesis **1985**, 645.
[119] B. Milenkov, M. Hesse, Helv.Chim.Acta **69**, 1323 (1986).
[120] B. M. Trost, T. R. Verhoeven, J.Am.Chem.Soc. **101**, 1595 (1979).
[121] V. G. Mairanovsky, Angew.Chem. **88**, 283 (1976), Angew.Chem. Int.Ed.Eng. **15**, 281 (1976).
[122] S. S. Kulp, M. J. McGee, J.Org.Chem. **48**, 4097 (1983).
[123] R.W. Freerksen, S. J. Selikson, R. R. Wroble, K. S. Kyler, D. S. Watt, J.Org.Chem. **48**, 4087 (1983).
[124] S. J. Selikson, D. S. Watt, J.Org.Chem. **40**, 267 (1975).
[125] R.W. Freerksen, D. S. Watt, Synth. Commun. **6**, 447 (1976).
[126] A. Donetti, O. Boniardi, A. Ezhaya, Synthesis **1980**, 1009.
[127] D. Savoia, E. Tagliavini, C. Trombini, A. Umani-Ronchi, J.Org.Chem. **45**, 3227 (1980).

[128] C. E. Berkoff, D. E. Rivard, D. Kirkpatrick, J. L. Ives, Synth. Commun. **10**, 939 (1980).

[129] M. Geier, M. Hesse, Synthesis **1990**, 56.

[130] D. Seebach, E. W. Colvin, F. Lehr, T. Weller, Chimia **33**, 1 (1979).

[131] J. E. McMurry, J. Melton, J. Org. Chem. **38**, 4367 (1973).

[132] N. Kornblum, A. S. Erickson, W. J. Kelly, B. Henggeler, J. Org. Chem. **47**, 4534 (1982).

[133] E. Keinan, Y. Mazur, J. Am. Chem. Soc. **99**, 3861 (1977).

[133] D. Seebach, E. W. Colvin, F. Lehr, T. Weller, Chimia **33**, 1 (1979).

[134] N. Ono, A. Kagi, Synthesis **1986**, 693.

VIII. Ring Expansion by Cleavage of the Zero Bridge in Bicycles

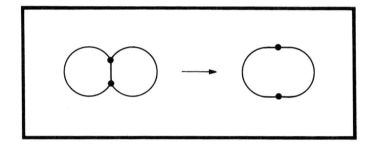

One of the three general ring enlargement methods mentioned in the introduction to this book is the cleavage of the shortest bridge in bicyclic systems. In most cases, the so-called zero bridge of a bicycle is broken. The zero bridge can be a single or a double bond, either between two carbon atoms, or a carbon and a nitrogen, or two nitrogen atoms. Depending on the nature and environment of this central bond, a large number of methods are known to cleave this bond. We have classified these reactions into three main groups. First, we will discuss fragmentation reactions leading to the ring expanded products. The second section deals with single bond cleavages of different kinds, and finally, we will consider the oxidative splitting of carbon, carbon double bonds.

VIII.1. Cleavage of the Zero Bridge in Bicycles by Fragmentation Reactions

3-Hydroxyketones can be generated by an aldol reaction, for example between two ketones. This reaction is reversible, that is, the 3-hydroxyketones can be transformed back into the two ketones by a carbon, carbon bond cleavage. Such an aldol system as part of a ring system is shown in the general structure **VIII/1**, Scheme VIII/1. The aldol is incorporated in the bicycle in such a way that the carbon, carbon bond, cleaved in a retro aldol reaction, is the zero bridge of the bicycle. A result of this reaction is the generation of an expanded monocyclic system, which we will discuss in more detail later (Chapter VIII.2).

If the carbonyl group in the bicyclic aldol system is reduced to an alcohol group, a diol, like **VIII/2**, will be formed. Replacement of the secondary alcohol group by a leaving group such as halogen, OTs, $NR_3^{\oplus}$ or $OH_2^{\oplus}$ leads to another ring system, which can be used for ring enlargement reactions.

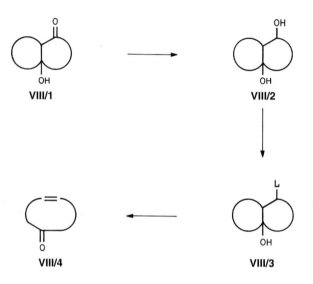

Scheme VIII/1.

The transformation of **VIII/3** to **VIII/4** is called a fragmentation[1) [3] [4]. As in the aldol reaction the reverse version of the fragmentation also is known (**VIII/4 → VIII/3**). An example of this reaction type is the so-called Prins reaction; the acid catalyzed (base catalysis is also possible) addition of an olefin to formaldehyde in order to get a 1,3-diol. Further examples are known in the field of transannular reactions in medium-sized rings [5].

In some cases, it appears that the mechanism of the fragmentation is E2, since an *anti* elimination has been observed [6]. A systematic investigation has been made on the behavior of the diastereoisomeric monotosylates **VIII/5**, **VIII/6**, **VIII/7**, and **VIII/8** under fragmentation conditions (KOtBu, HOtBu, 1h, 40°) [7], Scheme VIII/2. The results demonstrate the importance of the geometry of the substrate. In **VIII/6** and **VIII/7**, the bonds marked as a and b are antiperiplanar to each other with an angle of 180°. The reactions yielded the

---

1) This reaction type is called a Grob as well as a Wharton fragmentation [1] [2].

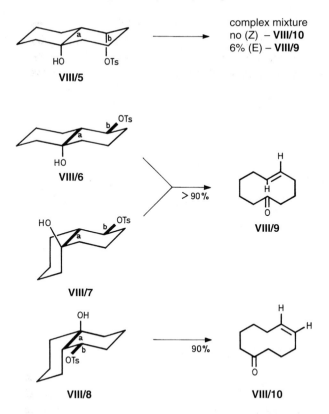

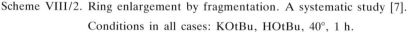

Scheme VIII/2. Ring enlargement by fragmentation. A systematic study [7].

Conditions in all cases: KOtBu, HOtBu, 40°, 1 h.

same product, (E)-**VIII/9**. A comparable geometry is present in compound **VIII/8**, which clearly gives the corresponding (Z)-isomer, **VIII/10**. The synclinal arrangement of a- and b-bonds in **VIII/5** does not favor a fragmentation reaction, and decomposition products are isolated, as well as unreacted starting material, **VIII/5**. Not more than 6 % of (E)-**VIII/9** was observed by gas chromatography. This amount might be expected from a non-concerted fragmentation *via* a carbocation; the other three appear to be formed by a concerted mechanism [7]. The process is general for 1,4-disubstituted systems; even dibromides can undergo elimination of bromine in the presence of zinc; **VIII/11** → **VIII/12** [8]. The disposition of the substituents determines the geometry of the olefinic bonds formed, allowing considerable control of the processes, as in reactions involving alkylborane fragmentation[2) **VIII/13** → **VIII/15**, **VIII/16** → **VIII/17** [10] [11] [12], Scheme VIII/3.

---

2) For a review on diene synthesis *via* boranate fragmentation, see ref. [9].

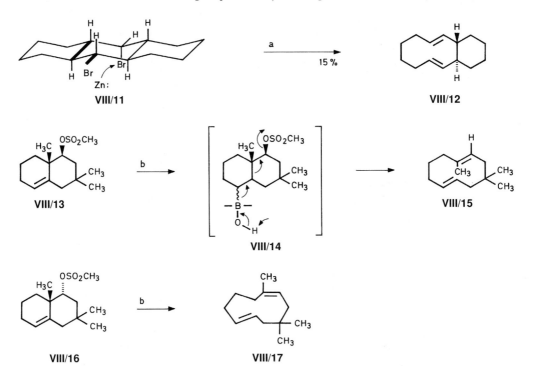

Scheme VIII/3. Fragmentations [9]. Before elimination the axial Br in **VIII/11** is converted to an equatorial ZnBr.

a) Zn, 1,2-dimethoxyethane   b) B_2H_6, THF-NaOH, H_2O.

The stereochemical aspects of the fragmentation reaction have been important in the syntheses of many complex molecules.

The stereospecific synthesis of (*E*)-6-methyl-5-cyclodecenone (**VIII/19**) was realized by treatment of the monomesylate **VIII/18** with potassium *tert*-butoxide [2], Scheme VIII/4.

The reaction principle was also applied to the synthesis of a precursor, **VIII/22**, of the sex excitant of the American cockroach *Periplanata americana*, periplanone-B (**VIII/20**) [13]. The bicycle, **VIII/21**, was synthesized in a four-step sequence. In **VIII/21** the bicyclic zero bridge between the two rings is anti-periplanar with respect to the leaving group at C(7) (see **VIII/21a**). Further-more, one of the orbitals of the alcoholate oxygen at C(1) has an antiperiplanar orientation to the zero bridge mentioned above. Compound **VIII/21** is first transformed into its dilithium salt and then, by adding trifluoromethanesulfonic anhydride, into the desired fragmentation product, **VIII/22** (in 44 % yield). A similar reaction was used to construct the nine-membered ketones **VIII/24** [14] and **VIII/26** [15]. As in example **VIII/21**, **VIII/23** (Scheme VIII/5) has one of the orbitals of the oxygen of the hydroxyl group antiperiplanar to bond a, and

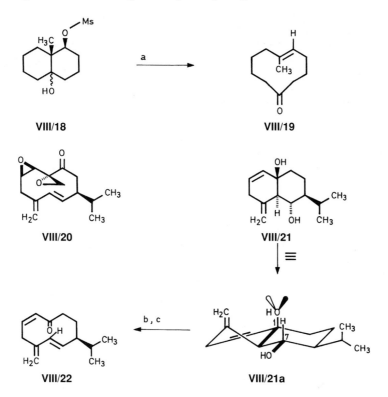

Scheme VIII/4. Ring enlargement by fragmentation. The boldface printed bonds in **VIII/21a** are antiperiplanar to each other.

a) KOtBu, HOtBu   b) 2 BuLi, −30°   c) 3 $(CF_3CO)_2O$, −20°.

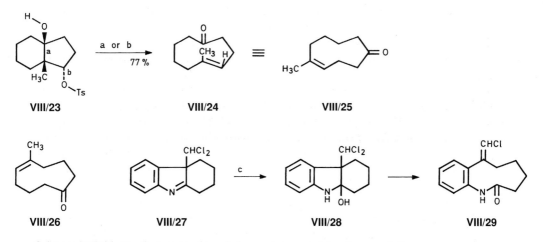

Scheme VIII/5. Further examples of ring enlargement by fragmentation.

a) NaH, THF, 35°, 24 h   b) KOtBu, THF, 20°, 20 h   c) NaOH.

bond a is antiperiplanar to bond b; an ideal situation for the fragmentation reaction. – Treatment of the carboline derivative, **VIII/27**, with NaOH/H$_2$O gives the aminoacetal, **VIII/28**. This yields, by fragmentation, **VIII/29**, (elimination of hydrochloric acid) [16]. Other examples are discussed in [17] [18] [19] [20] [21] [22] [23] [24].

Besides the fragmentation of the 6/6 and 6/5 annelated rings, a number of other ring combinations in bicycles have been investigated. A 6/3 system[3], Scheme VIII/6, has been used to explain the observation that the unexpected 6,7-dimethyl-3,4-benzotropolone (**VIII/33**) is formed when an acetic acid solution of **VIII/30** is treated with zinc powder [25][4].

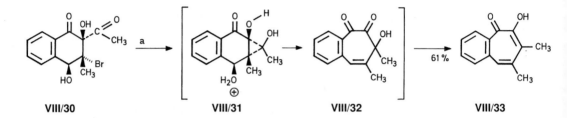

VIII/30   VIII/31   VIII/32   VIII/33

Scheme VIII/6. Reductive ring expansion by one carbon atom.

a) Zn, HOAc.

A 6/4 annelated ring system as part of a tricyclic intermediate, has been constructed to synthesize monocyclic (±)-phoracantholide M (**VIII/42**), Scheme VIII/7. The correct configurations at centers 1, 8, and 9 in **VIII/38**, are important for the fragmentation. These configurations are controlled by the intramolecular photo [2+2] cycloaddition of **VIII/36**. Borohydride reduction of the resulting ketone **VIII/37** is stereospecific under the influence of the two centers already formed. Because of its instability, the fragmentation product **VIII/40** could not be isolated. Instead, the central carbon, carbon double bond of the bicyclic **VIII/40** was oxidatively cleaved to give **VIII/41**, which was finally transformed into the desired **VIII/42** [26]. Similar oxidative ring expansion reactions, are discussed in Chapter VIII.3.

---

3) For other ring expansion reactions in which cyclopropanes are involved, see Chapter III.
4) Cyclopropanes are produced, if 3-bromoketones react with zinc in acetic acid (**VIII/30** → **VIII/31**). In the proposed intermediate **VIII/32** the hydroxy group can be lost by hydrogenolysis [25].

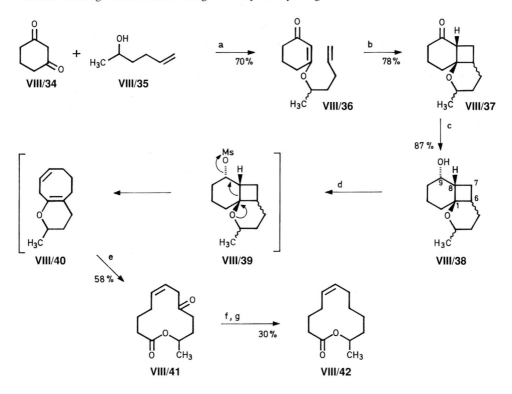

Scheme VIII/7. Synthesis of (±)-phoracantholide M (**VIII/42**) by two different types of ring enlargements [26].

a) TsOH   b) hν c) NaBH$_4$, CH$_3$OH
d) methanesulfonyl chloride, pyridine, CHCl$_3$, 0° → reflux
e) *meta*-chloroperoxybenzoic acid   f) TsNHNH$_2$, CH$_3$OH
g) [(C$_6$H$_5$)$_3$P]$_2$CuBH$_4$, CHCl$_3$, reflux.

In the course of a muscone (**VIII/48**) synthesis, the stereoelectronic conditions of a fragmentation in a 12/5 bicycle were carefully studied (Scheme VIII/8) [27]. Heating the epoxysulfone, **VIII/43**, with sodium amide gave only the hydroxysulfone **VIII/44**. The configuration of **VIII/44** was established by an X-ray analysis.

An equilibrium between the isomeric hydroxysulfones, **VIII/44** and **VIII/45**, was observed in the presence of butyllithium. Treatment of **VIII/44** with potassium *tert*-butoxide gave only the ring enlarged (*E*)-isomer **VIII/46**. The (*E*)-configuration of the double bond is the result of the stereoelectronic course of the fragmentation. Under the same reaction conditions the formation of the isomeric **VIII/47** from **VIII/45** was not observed; only compound **VIII/46** was isolated. This must happen *via* thermodynamically controlled epimerization of **VIII/45** → **VIII/44**. For similar reactions, see ref. [28].

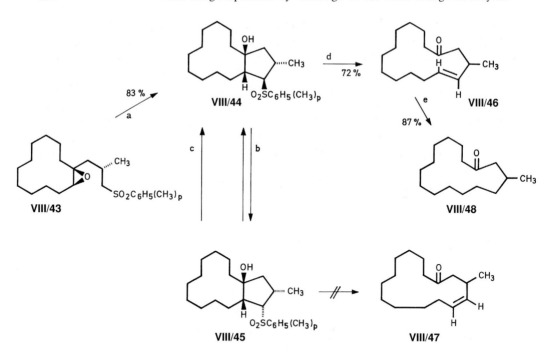

Scheme VIII/8. A synthesis of muscone (**VIII/48**) with a fragmentation as key step [27].

a) NaNH$_2$, toluene, reflux   b) BuLi   c) KOtBu
d) KOtBu, toluene, hexamethylphosphoramide, 120°, 15 h   e) H$_2$, Pd-C.

From the mechanistic point of view, it should be noted that the oxidative cleavage of a double bonded zero bridge in bicycles (compare Chapter VIII.3) might also be a fragmentation. The following may be an example of this phenomenon (Scheme VIII/9).

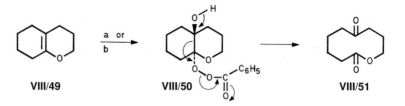

Scheme VIII/9. a) *meta*-Chloroperoxybenzoic acid   b) Formation of *trans*-diol followed by *meta*-chloroperoxybenzoic acid.

The decomposition of **VIII/50** of known configuration, prepared from **VIII/49**, gives to the ten-membered **VIII/51** by a fragmentation reaction [29] [30].

A different kind of fragmentation is observed when 15-pentadecanolide (**VIII/57**) is prepared from 2-oxocyclododecane-carbonitrile (**VIII/52**) [31] [32]. As shown in Scheme VIII/10, the methoiodide **VIII/53** is synthesized and afterwards converted to the 16-membered **VIII/56**. The fragmentation reaction is presumed to take place through hemiacetal intermediates such as **VIII/54** and **VIII/55**. However, despite many experiments, yields of **VIII/56** were never greater than 64 %, with the remainder being recovered as starting material. This was explained by argueing that both **VIII/54** and **VIII/55** were formed, but that only **VIII/54** had the correct antiperiplanar configuration for fragmentation, the **VIII/55** formed, presumably about 35 %, was then recovered as starting material. Epimerization between **VIII/54** and **VIII/55** can be excluded because other reactions were not observed [31].

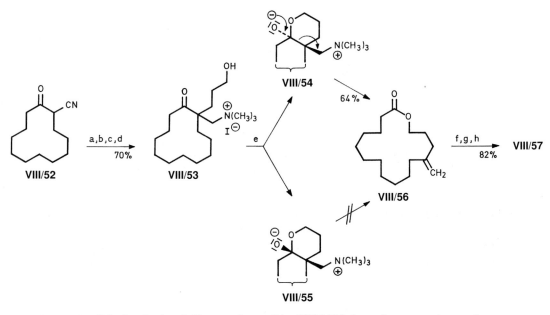

Scheme VIII/10. Synthesis of 15-pentadecanolide (**VIII/57**) by a fragmentation pathway [31] [32].

a) CH$_2$=CH-CHO b) NaBH$_4$ c) H$_2$-Pt, HCl d) CH$_3$I, CH$_3$OH, KHCO$_3$
e) NaH, dimethylformamide f) O$_3$, CH$_2$Cl$_2$ g) CH$_3$OH, TsNHNH$_2$
h) [(C$_6$H$_5$)$_3$P]$_2$CuBH$_4$, CHCl$_3$.

A fascinating synthesis of the twelve-membered lactone 5(*E*),8(*Z*)-6-methyl-5,8-undecadien-11-olide (**VIII/60**) is shown in Scheme VIII/11. The tricyclic system **VIII/58** and its isomer, **VIII/59**, starting materials for this reaction, were built up from three annelated six-membered rings. When both compounds were heated to their melting points, 180° and 220°, respectively, the evolution of carbondioxide and the formation of p-toluenesulfonic acid was observed [33]. In

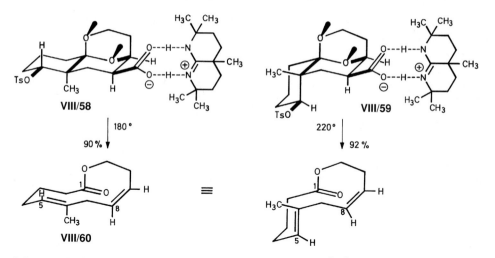

Scheme VIII/11. Double fragmentation in a tricyclic system [33]. Boldface bonds are antiperiplanar orientated.

both cases, two bonds inside the tricycle are broken during the formation of the monocycle of the enlarged system, **VIII/60**. The central carbon carbon bridge in both **VIII/58** and **VIII/59** is antiperiplanar to the equatorial tosyl group as well as to the electron pair orbitals of the acetal oxygens. The equatorial carboxylate, on the other hand, is antiperiplanar to the central acetal bond. This fragmentation process my be a one step reaction; the yields of **VIII/60** from both isomers are very high. A similar reaction is presented in ref. [34].

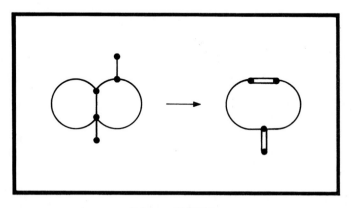

Scheme VIII/12.

Olefin bonds and ketones are formed by the fragmentation reactions discussed above (Scheme VIII/12). To get alkynes with ketones by similar processes, an alkenol instead of the alkanol must be present in the starting material.

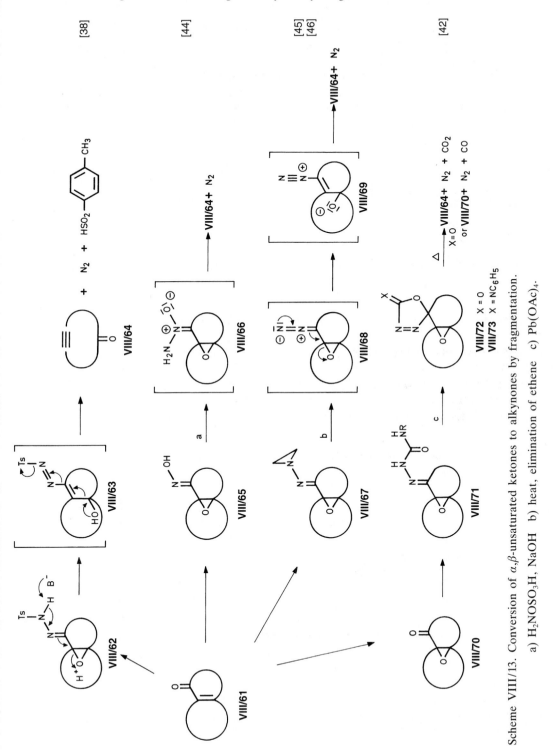

Scheme VIII/13. Conversion of α,β-unsaturated ketones to alkynones by fragmentation.

a) H₂NOSO₃H, NaOH   b) heat, elimination of ethene   c) Pb(OAc)₄.

These reactions become more realistic if derivatives of α,β-epoxyketones are used. These can be easily prepared by oxidation of the corresponding α,β-unsaturated ketones with peracids or hydrogen peroxide (Scheme VIII/13). The driving force in studying this kind of reactions is the importance of the economic synthesis of the natural 15-membered ketones, muscone and cyclopentadecanone, from cyclododecanone, an easy available and inexpensive starting material (Scheme VIII/14) [35] [36].

A number of different derivatives of the α,β-epoxyketone may be used in the key fragmentation reaction, see Scheme VIII/13. The reaction was discovered using the tosylhydrazone **VIII/62**[5)] [37] [38] [39] [40] [41], which gave, under acidic conditions (*e.g.* CH_2Cl_2/HOAc) and low temperatures the cycloalkynone, **VIII/64**, as well as nitrogen and a sulfinic acid. Improvements of this reaction are shown in ref. [42]. In a second version, the α,β-epoxyketonoxime, **VIII/65**, is treated with hydroxylamine-O-sulfonic acid in the presence of base to form the same cycloalkynone, **VIII/64**, and nitrogen [44]. In a third approach, the α,β-epoxyketones are transformed to the hydrazones of N-aminoaziridines such as **VIII/67**, which undergo the fragmentation thermally to give **VIII/64**, as well as ethene and nitrogen [45] [46]. A final possibility is also shown in Scheme VIII/13. The compound, **VIII/61**, was transformed into its semicarbazone, **VIII/71**, which, by lead tetraacetate oxidation, was converted into 5,5-disubstituted 2-imino-Δ^3-1,3,4-oxadiazolines, **VIII/73**. The corresponding oxadiazoline, **VIII/72**, is formed by aqueous acid hydrolysis of **VIII/73** and is usually a mixture of diastereoisomers. Thermolysis in refluxing acetonitrile, yields cycloalkynone, **VIII/64**, nitrogen, and carbondioxide or the starting α,β-epoxyketone, **VIII/70**, nitrogen, and carbonmonoxide. The two products, **VIII/64** and **VIII/70**, are a consequence of competitive decomposition *via* two pathways. Which pathway predominates may be controlled partly by the polarity of the solvent [42].

A ($\pm$)-muscone (**VIII/48**) synthesis is given in Scheme VIII/14. In this reaction the key step is an application of the α,β-epoxyketone $\rightarrow$ alkynone fragmentation. Beside the way given in this Scheme, alternatives have been worked out for the synthesis of the bicyclic intermediate, see ref. [47] [48].

A few observations can be made concerning the synthetic application of this fragmentation reaction. The hydroxylamine version seems to take place even in sterical hindered α,β-epoxyketones. Precise reaction conditions, are also necessary to make the procedure successful [38].

Because of preparative difficulties in the epoxidation of α,β-unsaturated ketones (influenced by steric factors, *e.g.* CH_3-group in **VIII/76** [38]), a direct conversion of these compounds was developed [43].

Aldehydes and ketones can be regenerated from their tosylhydrazones by treatment with N-bromosuccinimide/CH_3OH/acetone [49] or with H_2O_2/K_2CO_3 [50]. The mechanism of the first conversion is given in Scheme VIII/15.

---

5) The α,β-epoxyketone $\rightarrow$ alkynone fragmentation is called Eschenmoser fragmentation [43].

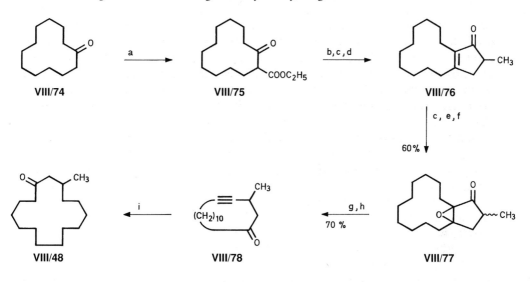

Scheme VIII/14. Synthesis of (±)-muscone (**VIII/48**) using the tosylhydrazone approach of the α,β-epoxyketone $\rightarrow$ alkynone fragmentation [38].

a) $CO(OC_2H_5)_2$, NaH b) $H_2C=C(CH_3)COOCH_3$ c) $NaBH_4$
d) polyphosphoric acid e) $C_6H_5CO_3H$ f) CrO_3
g) $TsNHNH_2$, CH_3OH, 4° h) acetone, heat i) H_2/Pd-C.

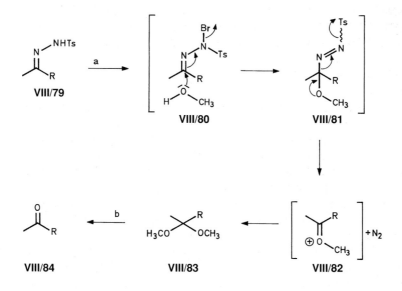

Scheme VIII/15. Cleavage of tosylhydrazones after N-bromosuccinimide treatment [43] [49].

a) N-Bromosuccinimide, CH_3OH, acetone b) $H_3O^{\oplus}$.

Vinylogous hydrazones should react in a similar manner. If a 1,4-attack takes place in an appropriate system (Scheme VIII/16) a ring enlargement would be expected. The success of such a fragmentation depends upon whether the reaction can be directed in favor of the 1,4-addition of the nucleophile [43]. Several reaction conditions have been studied, but 1,2-addition cannot be completely excluded. On the other hand, 1,2-addition is not a serious problem because starting material, **VIII/85**, is regenerated, see Scheme VIII/16. Depending on reaction conditions, the product ratio, **VIII/89**:**VIII/85**, varies between 1:1 and 7:1 (R = CH₃); in the corresponding cyclopentadecanone series, the ratio of the comparable products (R = H, without the methyl group) is between 1.7:1 and 19:1 [43].

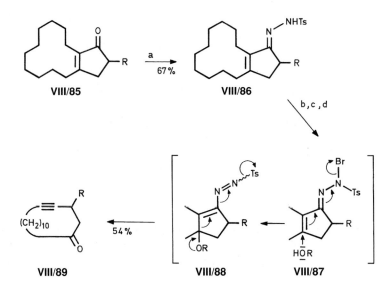

Scheme VIII/16. Tosylhydrazone fragmentation [43].

R = H or CH₃
a) TsNHNH₂, C₂H₅OH   b) (CH₂OH)₂/THF 1:2
c) N-bromosuccinimide, acetone, 15°/3 min
d) NaHSO₃, H₂O, 55°, 30 min.

The reaction sequence made it possible to investigate the chemistry of cycloalkynones. One of the smallest rings which has been synthesized is the nine-membered **VIII/93**[6] [51]. Under acid catalysis, **VIII/93** can be converted back into the starting material, **VIII/90**, Scheme VIII/17. The sequence **VIII/90** →

---

6) 5-Cyclononynone (**VIII/93**) shows no IR absorption (neat) for C≡C because of the symmetry of the molecule (1695cm⁻¹ for C=O) [51].

VIII/91 → **VIII/92** → **VIII/93** → **VIII/90** is like a chemical *perpetuum mobile* (perpetual motion). However, after one sequence the amount of **VIII/90** is reduced to approximately 40 %!

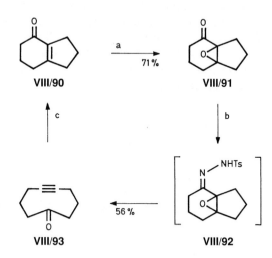

Scheme VIII/17. A chemical "*perpetuum mobile*" (perpetual motion) [51].

a) H_2O_2, CH_3OH, KOH   b) $TsNHNH_2$, CH_2Cl_2, HOAc
c) 2N H_2SO_4, H_2O, C_2H_5OH.

Further examples are the synthesis of 5-cyclodecynone [52][53] and the fragmentation of 1,2-epoxy-3-diazirine-5α-androstan-17β-ol by treatment with sodium iodide and acetic acid. (The A ring is opened between C(2) and C(3) to give the 1-oxo-2,3-alkyne derivative) [54].

VIII.2. Cleavage of Zero Bridged Single Bonds in Bicycles

Reduction and Hydrolysis of Cyclic Diaminoacetals and Aminoacetals

When structural elucidations of natural products of low molecular weight were done by chemical methods, unexpected transannular reactions occasionally made such work extremely difficult. Phenomena were observed, which could only be explained by reactions of functional groups in a ring with one another producing ring enlargement or ring contraction reactions, or, in some cases, by equilibria between open and closed systems. Such transannular reactions depend on the reaction conditions as well as the structures of the substrates.

For example, large rings containing amino and ketone groups are often responsible for this phenomenon in the alkaloid chemistry. A few examples will be discussed below.

The application of the Emde degradation conditions (0.1 N KOH/C_2H_5OH,
H_2/Pt) to the quaternary curare alkaloid, mavacurine iodide (**VIII/94**), led to
the so-called ε_2-dihydromavacurine (**VIII/95**), a tertiary base, Scheme III/18.
Protonation (*e.g.* HCl/H_2O) or methylation (CH_3I) of **VIII/95** led to the transannular reaction products, **VIII/96** and **VIII/97**, respectively. The quaternary
compound, **VIII/96**, returns to tertiary **VIII/95** (reversible Hofmann elimination) in the presence of base or on tlc (silica gel) [55].

The alkaloid tubifolidine (**VIII/98**) is isomerized to condyfoline (**VIII/100**)
and *vice versa* in a sealed tube at 120° without solvent [56], Scheme VIII/18.

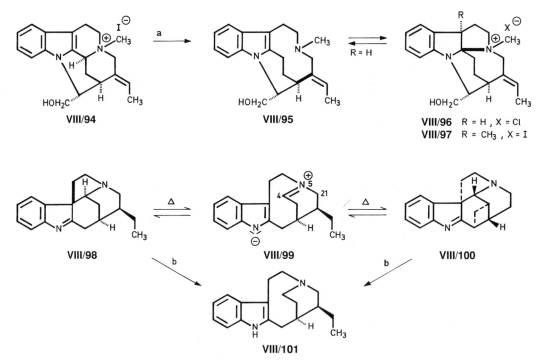

Scheme VIII/18. Examples of ring enlargement reactions in alkaloid chemistry.

a) H_2/Pt, KOH, C_2H_5OH b) KBH_4, CH_3OH.

The intermediates in this reaction are the nine-membered **VIII/99** and its isomer
with a C(5),C(21) double bond. Reductions of alkaloids, **VIII/98** and **VIII/100**
and intermediate **VIII/99**, with potassium borohydride in methanol, gave the
same compound, **VIII/101**, with a medium sized ring.

Depending on the reaction medium some cyclic alkaloids behave as ring enlarged or as ring contracted compounds. In Scheme VIII/19 indole alkaloids vomicine and perivine, the isoquinoline alkaloid protopine, and pyrrolizidines are given with their ring contracted forms.

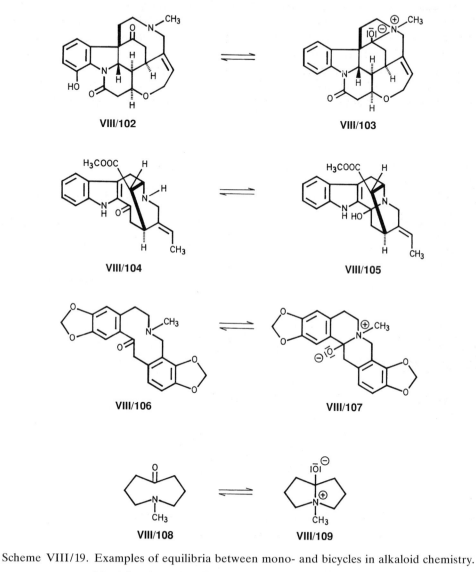

Scheme VIII/19. Examples of equilibria between mono- and bicycles in alkaloid chemistry.

Vomicine from *Strychnos nux-vomica* L. [57]
Perivine from *Catharanthus* and *Gabunia* species [57]
Protopine is widespread in the plant families of *e.g.* Fumariaceae, Hypecoaceae, Nandinaceae, and Papaveraceae [58]
Pyrrolizidine alkaloids [58] [59]

The Hofmann elimination can be used for ring enlargement as shown in Scheme VIII/18. A recent example, **VIII/110** → **VIII/111**, is given in Scheme VIII/20 [60]. The new double bond in **VIII/111** allows the formation of a conjugated system, consisting of an isoxazole and a benzene ring.

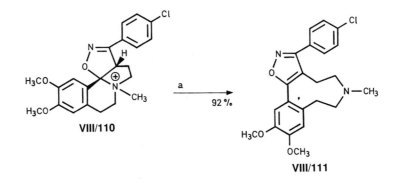

Scheme VIII/20. Use of the Hofmann elimination reaction for ring enlargement.

a) CH₃ONa, CH₃OH.

Another example of a spontaneous ring enlargement is shown in Scheme VIII/21. The pentacycle, **VIII/112**, in chloroform solution, was placed on a chromatography column (silica gel; water slurry) for 18 hours and was then eluted. Unchanged starting material (10 %) and 90 % of the isomeric tetra-cyclic, **VIII/113**, were obtained [61]. Based on the observations of the authors, both annelated benzene rings were necessary for the ring expansion reaction. (Presumably, HCl from the chloroform caused the formation of the nine-membered ring in **VIII/113**).

Hydrolyses of the bridged diaminoacetals such as **VIII/116**, prepared by the reaction of azirines with different reagents (see Chapter III) lead to ring enlarged products of type **VIII/117** [62]; further examples are mentioned in ref. [63] [64].

The reduction of substituted 4-hydroxy-5,6-dihydropyrimidins such as **VIII/114** is a reaction used several times as key step in the syntheses of polyamine alkaloids, Scheme VIII/21. In the presence of NaCNBH₃/AcOH at 50°, ring enlarged azalactams of type **VIII/115** are obtained in yields of about 90 %. Azalactams, prepared by this method, are nine- [65], thirteen- [66], and seventeen-membered [67] [68] [69].

The zero-bond in a bicyclic system with certain structural features can be cleaved by the von Braun degradation. In this case a nitrogen atom must be in a bridgehead position. For example, ring cleavage of the dihydroindole derivative, **VIII/118**, gives benzazocines **VIII/119** and **VIII/120** in good yields [70] [71], Scheme VIII/22.

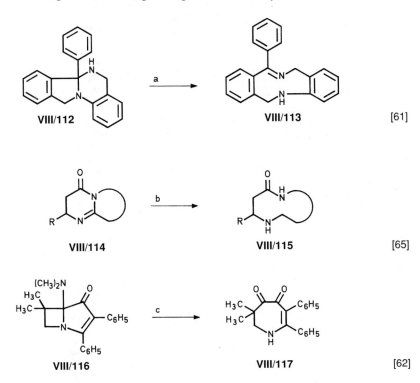

Scheme VIII/21. a) SiO$_2$, CHCl$_3$ b) NaCNBH$_3$, AcOH   c) 2 N HCl, H$_2$O.
      R=C$_6$H$_5$, Alkyl

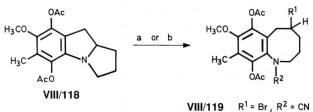

Scheme VIII/22. a) BrCN, C$_6$H$_6$, 20°, 15 h → **VIII/119**, 56 %
        b) (CF$_3$CO)$_2$O, sealed tube, 150–160°, 1.5 h → **VIII/20**, 79 %.

Reduction of Hydrazines

The use of hydrazine derivatives for ring enlargement reactions has not been studied very carefully. The idea would be to prepare bicyclic hydrazines or hydrazides, in which both nitrogen atoms of the hydrazine occupy bridgehead positions. The reductive cleavage of the nitrogen, nitrogen bond can then be carried out by catalytic hydrogenation (Raney-Ni [72]), by treatment with sodium in liquid ammonia, or with sodium naphthalenide in 1,2-dimethoxy-ethane [73]. Probably, the first compound prepared by this kind of ring enlarge-ment reaction was 1,5-diazacyclooctane, formed as a by-product in the synthesis of 1,2-diazabicylo[3.3.0]octane [74]. The first actual investigation of this reac-tion was undertaken in the course of the syntheses of 1,5-diazacyclononane and 1,6-diazacyclodecane. Both compounds were prepared in good yields [72]. In a typical procedure, ethyl acrylate and hydrazine hydrate were heated in a molar ratio of 1:2 to give 1,5-diazabicyclo[3.3.0]octane-4,8-dione in 80 % yield. This compound yielded on reduction, first with LiAlH$_4$ (81 % yield) and afterwards with H$_2$/Raney-Ni (75 %) 1,5-diazacyclooctane [75]. This sequence is an elegant way to synthesize medium sized diazacycloalkanes.

In a similar reaction, the reduction of 3,7-dimethyl-1,5-diazabicyclo[3.3.0]-octane-2,6-dione (**VIII/127**) to the cyclooctane derivative **VIII/128** (Scheme VIII/23) was nearly quantitative.

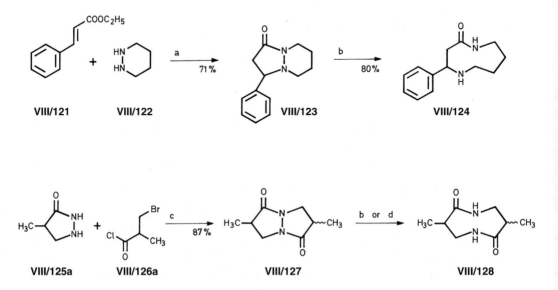

Scheme VIII/23. Reduction of hydrazine derivatives as a method of ring enlargement.

a) Heat, 27 h   b) Na, liq. NH$_3$ c) CH$_2$Cl$_2$, 0°
d) Na, naphthalene, 1,2-dimethoxyethane.

This reaction was applied to the synthesis of the macrocyclic polyamine alkaloid celacinnine [65]; a key step of this synthesis was the conversion of the bicyclic **VIII/123** to nine-membered **VIII/124**, see Scheme VIII/23.

The twelve-membered cyclo-dipeptide glidobamine will be synthesized using the reductive cleavage of the nitrogen, nitrogen bond in a hydrazine derivative as a key step [76].

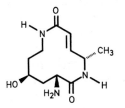

Glidobamine

The Retro Mannich and the Retro Aldol Reaction

As mentioned at the beginning of this chapter the retro aldol and the retro Mannich reaction can be used for ring expansion of proper substituted bicyclic systems.

An example of the retro Mannich type reaction is the formation of the 14-membered ring compound **VIII/126a** from the tetrahydroisoquinoline derivative **VIII/125a** in 11 % yield, Scheme VIII/24 [77].

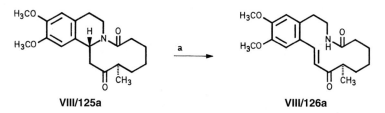

VIII/125a   VIII/126a

Scheme VIII/24. Examples of retro Mannich reaction in ring enlargement.

a) $NaOC_2H_5$, C_2H_5OH.

The general use of the retro aldol reaction for ring expansion is limited because of the difficulty of preparing properly substituted starting material. One interesting synthetic approach is the photochemical cycloaddition of an enol acetate and a cycloalkene (shown in Scheme VIII/25) [78]. Irradiation of cyclopentene (**VIII/127a**) and 3-acetoxy-5,5-dimethyl-2-cyclohexenone (**VIII/128a**) gave the two isomeric 2-acetoxy-4,4-dimethyl-tricyclo[6.3.0.0²,⁷]undecan-6-ones, **VIII/129** and **VIII/130**, together in 65 % yield. Both isomers behave similarly under basic or acidic reaction conditions. In sodium methoxide, only the α,β-unsaturated tricyclic ketone can be observed (by loss of acetic acid). But at room temperature after two weeks (!), in the presence of CH_3OH/H_2SO_4, **VIII/129** and **VIII/130** were converted to the eight-membered diketone, **VIII/131**, isolated in a yield of about 50 %. Because of the equilibrium, the retro aldol reaction in general must be performed under the same reaction conditions used to prepare the aldol itself. Depending on the ring size of the retro aldol products, "trivial" transannular reactions[7] can be observed. Occasionally they can lead to side products.

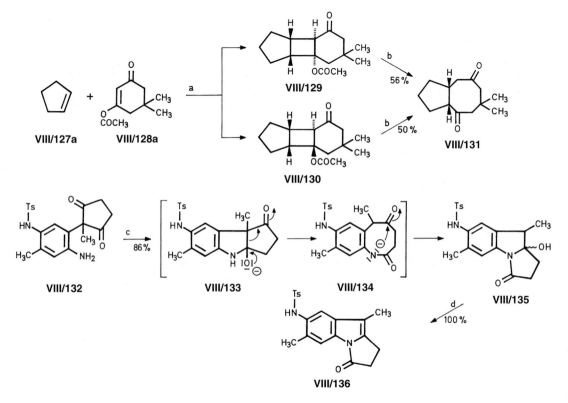

Scheme VIII/25. The retro aldol approach for ring enlargement.

a) hν b) H_2SO_4, CH_3OH, 20°, 14 d   c) NaH, THF, 20°,   d) HOAc, 20°.

---

7) The intramolecular base- or acid-catalyzed condensation of a ring compound containing carbonyl groups and adjacent active methylene groups, is a standard technique of organic chemistry. Prelog suggested that such reactions should be termed "trivial" transannular reactions in contrast with reactions, which are unique to medium-sized rings. An example of such a trivial reaction is the base-catalyzed transformation of cyclodecane-1,6-dione into bicyclo[5.3.0]decan-1[7]-en-2-one [81], while the formation of bicyclo[4.4.0]decan-1-en-2-one from cyclodecanone by treatment with two moles of N-bromosuccinimide [5] [82] is a model for the non trivial one.

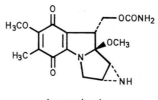

mitomycin A

In the case of synthetic studies on mitomycins (*e.g.* mitomycin A), both techniques were applied to synthesize model compounds. A retro aldol reaction, followed by a trivial transannular reaction, was performed in the same pot and under the same conditions (sodium hydride) [79] [80], Scheme VIII/25. Thus, compound **VIII/132** gave the ring closed intermediate, **VIII/133**, which by a retro aldol reaction, yielded the eight-membered intermediate, **VIII/134**. A transannular reaction in **VIII/134** gave the acid labile **VIII/135**, which led finally to the indole derivative, **VIII/136**.

In a series of experiments, retro aldol products such as **VIII/140** were obtained, when dihydroisoquinoline **VIII/137** was treated with a nonenolizable β-diketone, **VIII/138**, see Scheme VIII/26 [83].

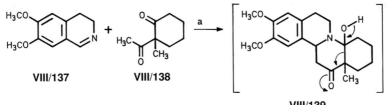

35 %

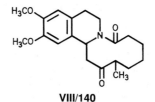

Scheme VIII/26. Retro aldol reactions.

a) H₂O, 20 h reflux.

VIII.3. Cleavage of the Zero Bridge in Bicycles by
Retro Diels-Alder Reaction

Retro Diels-Alder reactions in bicyclic systems or Diels-Alder reactions in corresponding monocyclic systems are of great interest. If, in bicyclic systems, the
central bridge is involved in the retro reaction, the process will lead to a ring
enlargement. Compounds which isomerize by this type of reaction belong to the
group of molecules, which show valence tautomerization. Two examples
(**VIII/141** ⇄ **VIII/142** and **VIII/143** ⇄ **VIII/144**) of many are given in Scheme
VIII/27. The reaction of diazomethane with benzene under irradiation with
light (UV or sun light) results in cycloheptatriene, a valence isomerized bicycle.

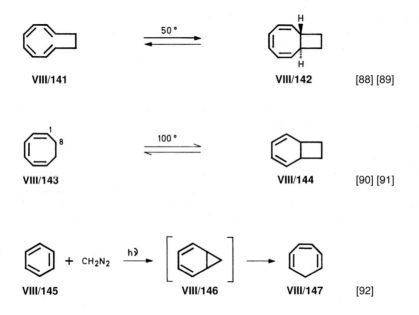

Scheme VIII/27. Valence tautomerization.

A series of substituted 1,3,5-cyclooctatrienes (*e.g.* **VIII/150**) has been synthesized according to Scheme VIII/28. They generally exist in equilibrium with
their valence tautomers, bicyclo[4.2.0]octa-2,4-dienes. The equilibrium is lar-
gely effected by the nature and position of the substituents. They were isolated
as the sole valence tautomers. This fact indicates the stabilization provided by
the conjugation with the carbonyl group, is strong enough to maintain a 1,3,5-
cyclooctatriene structure [84], Scheme VIII/28.

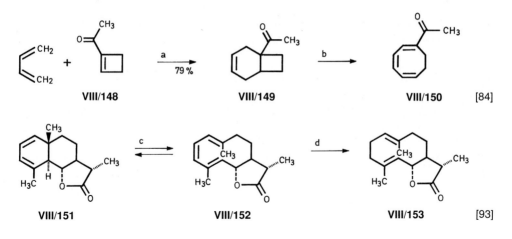

Scheme VIII/28. a) AlCl₃, CH₂Cl₂
   b) N-bromosuccinimide, CCl₄ Li₂CO₃, LiF, powdered soft glass, hexa-
   methyl-phosphoramide
   c) hν, CH₃OH, 18° d) H₂, Raney-Ni, CH₃OH.

For similar reactions, see ref. [85] [86]. The ratio of the components in the equilibrium **VIII/143:VIII/144** depends very much on the number of substituents. The percentage of the bicycle is increased with substitution (*e.g.* no substituents, 10.8 % bicycle; 7,7-dimethoxy, >95 %) [87].

Irradiation of the tricycle **VIII/151** is carried out in methanol solution at 18° under an atmosphere of argon, using a low-pressure mercury discharge tube. The product of this reaction was triene **VIII/152**. After about 1 hour of irradiation, a photostationary state was reached. Because **VIII/152** was found to be thermally unstable, a selective reduction was accomplished by bubbling hydrogen through a cold (−18°) methanolic solution of the photolysis products after addition of Raney-nickel. To receive compound **VIII/153**, dihydrocostunolide, was the goal of this total synthesis [93]. Another synthesis of **VIII/153** was achieved later [94].

VIII. 4. Oxidative Cleavage of the Zero-Ene-Bridge in Bicycles

Several medium and large ring compounds isolated from natural sources contain a ketone or a lactone group. Such molecules might be prepared by splitting, oxydatively, a zero bridged double bond in a bicycle[8]. The double bonded bicyclic system must be easily synthesized and the oxidation product, usually

---

8) For mechanistic considerations see Chapter VIII.1.

having two rather than one carbonyl groups, should be capable of being specifically transformed into the desired product.

Muscone (**VIII/48**) is one of the "most attractive" large ring compounds of natural origin for synthetic chemists. One of the approaches to its synthesis contains the oxidative cleavage of a zero bridged double bond [85]. The starting material was cyclododecanone (**VIII/74**, Scheme VIII/29), which was converted to 1-(trimethylsilyl)cyclododecene (**VIII/157**) by treatment with benzenesulfonylhydrazide. The cyclododecanone-benzenesulfonylhydrazone (**VIII/154**) was first converted to the dianion, **VIII/155**, by treatment with BuLi at −45°. The dianion on warming to −30°, spontaneously decomposed to the vinyl carbanion, **VIII/156**, nitrogen and benzenesulfinate. Species **VIII/156** was trapped

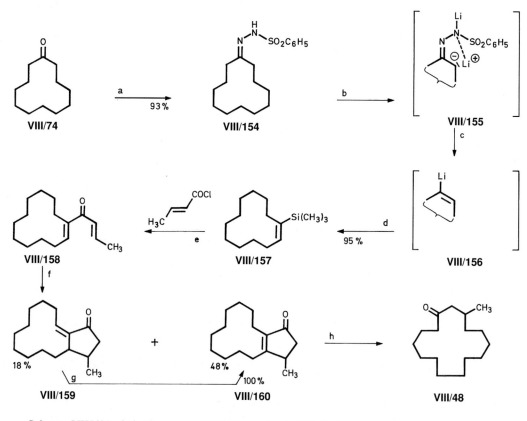

Scheme VIII/29. (±)-Muscone (**VIII/48**) synthesis [85] [95].

a) C₆H₅SO₂NHNH₂
b) C₄H₉Li, hexane, N,N,N′,N′-tetramethylethylenediamine, −45°
c) −30°, loss of N₂ and C₆H₅SO₂Li   d) ClSi(CH₃)₃
e) AlCl₃, CH₂Cl₂, 25°   f) BF₃·(C₂H₅)₂O, C₆H₅, reflux
g) RhCl₃, C₂H₅OH   h) 1. NaBH₄   2. O₃   3. N₂H₄, KOH   4. CrO₃.

with chlorotrimethylsilane to generate **VIII/157**. The next step was an annela-
tion of cyclopentenone. First, a Friedel-Crafts acylation led to the regiospecific
product **VIII/158**. The cyclization step was performed with borontrifluoride
etherate in hot benzene. The two cyclopentenone derivatives **VIII/159** and
VIII/160 (Scheme VIII/29) were then quantitatively isomerized to **VIII/160**.

The cleavage of the double bonded zero bridge was performed with ozone.
In order to keep the carbonyl group in compound **VIII/160** for the final product
VIII/48, this carbonyl group was reduced to the alcohol. The two carbonyl
groups, generated by ozone, were reduced under Wolff-Kishner conditions,
and, finally, the CrO_3 oxidation of the alcohol gave muscone (**VIII/48**) [85]. The
synthesis of muscone, outlined in Scheme VIII/29, is an alternative route for the
earlier described conversion of cyclododecanone to muscone [96]. Cyclopentadecanone (**VIII/163**, exaltone®) has been synthesized from cyclododecanone
(**VIII/74**) in a comparable, but yet interesting way [28]. The transformation
of compound **VIII/74** to bicyclo[10.3.0]pentadec-1(12)en-13-one (**VIII/85**,
Scheme VIII/30) was carried out by classical methods [48]. The reduction of
the α,β-unsaturated ketone with Raney-nickel (H_2/1% NaOH-MeOH) gave
two isomeric saturated alcohols, which were dehydrated ($C_6H_5SO_3H$, toluene)

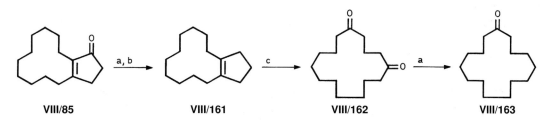

VIII/85   VIII/161   VIII/162   VIII/163

Scheme VIII/30. A synthetic pathway to cyclopentadecanone (**VIII/163**) [28].

a) H_2, Raney-Ni   b) $C_6H_5SO_3H$ c) O_3, CH_2Cl_2.

and isomerized to **VIII/161**, Scheme VIII/30. Ozonisation of the double bond
yielded the diketone **VIII/162**, and a partial catalytic hydrogenation (alkaline
solution, Raney-nickel) led to cyclopentadecanone (**VIII/163**) [28]. Similar
oxidative cleavages are mentioned in ref. [51]. Most of the products constructed
by the oxidative double bonded zero bridge splitting, are lactones or even
lactams. This preference is due to the fact that oxidation of the double bond
yields two carbonyl groups. If one of the carbonyl groups is part of a lactone or
lactam, the second one can be specifically reduced or transformed into other
functional groups. Two further examples are given in Scheme VIII/31.

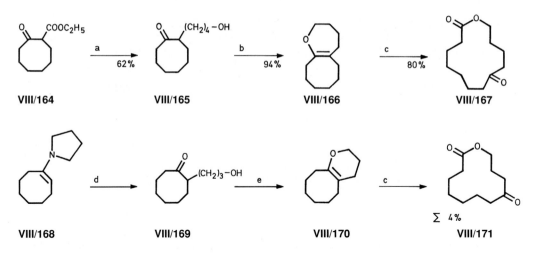

VIII/164   **VIII/165**   **VIII/166**   **VIII/167**

VIII/168   **VIII/169**   **VIII/170**   Σ 4% **VIII/171**

Scheme VIII/31. Oxidative cleavage of double bond zero bridge in bicycles [97] [98].

a) NaH, Br-$(CH_2)_4$-OCOCH$_3$ – KOH, H$_2$O
b) slow distillation *in vacuo*   c) *meta*-chloroperoxybenzoic acid, CH$_2$Cl$_2$
d) CH$_2$=CH-COOC$_2$H$_5$, C$_2$H$_5$OH – LiAlH$_4$ acid   e) TsOH, C$_6$H$_6$.

The incorporation of a ω-functionalized side chain in the α-position of ethyl 2-oxocyclooctane-1-carboxylate (**VIII/164**) yielded **VIII/165**, which by loss of water (distillation *in vacuo*) gave 9-oxabicyclo[6.5.0]tridec-1(8)-ene (**VIII/166**). Oxidation of **VIII/166** with an excess of *meta*-chloroperoxybenzoic acid for a short period of time (to avoid Baeyer-Villiger oxidation of the product oxolactone to dilactones[9)]) gave a 46% yield (from **VIII/164** of **VIII/167**) [98].

An alternative procedure, starting with the enamine **VIII/168**, gave the twelve-membered oxolactone **VIII/171** *via* **VIII/169** and **VIII/170** in a yield of 4% (from **VIII/168**, Scheme VIII/31), [97]. Syntheses of macrocyclic lactones with an annelated aromatic ring are described by various authors. Oxidations of compounds of type **VIII/172** have been investigated in order to prepare aromatic oxolactones of type **VIII/173**, [99], Scheme VIII/32[10)].

---

9) Alternative oxidation reagents, *tert*-butylhydroperoxide, molybdenum hexacarbonyl, or lead tetraacetate oxidation of the corresponding glycol, were tried in order to avoid dilactone formation, but the yields were not satisfactory [98].

10) The treatment of cyclic enol ethers with alkyl nitrites or with nitrosyl chloride gave oximino macrolides in almost quantitative yield [100] [102]. The furan derivatives are inert to hydrolytic nitrosation [100].

n = 1 – 4,8   a) C$_4$H$_9$ONO, C$_2$H$_5$OH, H$_2$O.

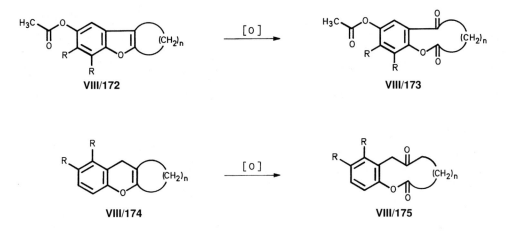

Scheme VIII/32. Synthesis of macrocyclic lactones with an annelated aromatic ring by ring enlargement [100] [101].

R = H or CH=CH-CH=CH- n = 4, 5, 6, 10

Meta-chloroperoxybenzoic acid or osmiumtetroxide/sodium periodate chromic acid anhydride in acetic acid or acetic anhydride give good results with the benzofurans, but was unsatisfactory in the naphthofuran (**VIII/172**, R-R, CH= CH-CH=CH-) series. Ozonolysis, however, was very effective with both types of compounds (yields 63–86 %).

Analogous reactions have been performed in similar benzo- and naphthopyrane series [101]. Oxidation of benzopyrane derivatives **VIII/174** with chromic acid/acetic acid, chromic acid anhydride/acetic acid anhydride, Jones reagent, or ozone gave only complex mixtures of products. Only *meta*-chloroperoxybenzoic acid in dichloromethane was successful (**VIII/175** as naphthoketolactones with n = 4, 5, 6 in 70, 49, and 60 % yield).

The preparation of a number of medium ring benzoic acid lactones was achieved by treatment of compounds such as **VIII/176** with an excess of *meta*-chloroperoxybenzoic acid in dichloromethane, Scheme VIII/33 [103]. However, this oxidation reaction is not general for the synthesis of aromatic lactones. If the same reaction conditions are used as in the conversion of **VIII/176** to **VIII/177**, the methoxy derivative **VIII/178** is not transformed into the corresponding lactone. Instead the cyclic carbonate **VIII/183** was isolated in a yield of 50 %. The proposed mechanism of this abnormal reaction is shown in Scheme VIII/33. From model compounds, the methoxyl group in the *para*-position to the center of oxidation seems to be important for the formation of **VIII/183** [103]. The carbonate **VIII/183** is unstable in aqueous alkaline medium and decomposes to the spiro compound, **VIII/185**, Scheme VIII/33 [103]. For an analogous reaction, see ref. [104].

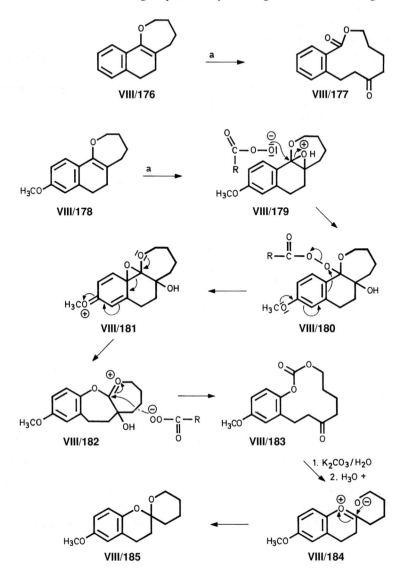

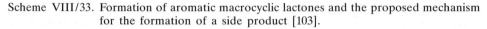

Scheme VIII/33. Formation of aromatic macrocyclic lactones and the proposed mechanism
for the formation of a side product [103].

a) *Meta*-chloroperoxybenzoic acid, CH_2Cl_2 b) K_2CO_3, $H_2O - H_3O^\oplus$.

Oxidation of cyclic enamins or unsaturated lactams, corresponding to the cyclic enol ethers, led to oxolactams and oxoimides, respectively [105].

The synthesis of the naturally occurring dihydrorecifeiolide (**VIII/191**)[11)] was realized by an oxidative cleavage of the double bond zero bridged bicycle as the key step, Scheme VIII/34. Methylation of the aldehyde, **VIII/187**, prepared from the cyclooctanone derivative, **VIII/186**, by a Michael reaction, gave the secondary alcohol, **VIII/188**, in high yield, when the reaction was carried out with dimethyltitaniumdiisopropoxide. However, instead of the expected ring enlargement (compare Chapter VII), deethoxycarbonylation [106] [107] [108] [109] took place under the influence of the fluoride ion. The bicycle **VIII/189** was finally formed by distillation. Oxidation of the central double bond with

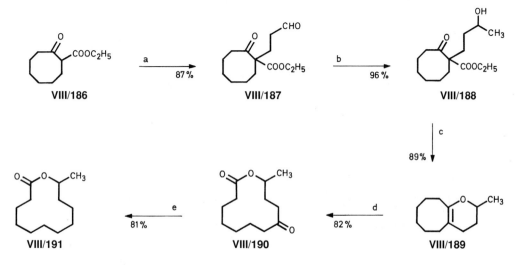

Scheme VIII/34. Synthesis of (±)-dihydrorecifeiolide (**VIII/191**) [110].

> a) CH₂=CH-CHO, Bu₃P   b) (CH₃)₂Ti(O-iPr)₂, THF
> c) Bu₄NF, THF – distillation   d) *meta*-chloroperoxybenzoic acid, CH₂Cl₂
> e) TsNHNH₂, CH₃OH-[(C₆H₅)₃P]₂CuBH₄, CHCl₃.

meta-chloroperoxybenzoic acid in the presence of potassium fluoride gave the best results, [112]. The formation of the corresponding tosylhydrazone and its reduction with bis(triphenylphosphine)copper(I)-tetrahydroborate resulted in the desired lactone, **VIII/191**, in 81 % yield (49 % from **VIII/186**) [110]. The synthesis [26] of (±)-phoracantholide M (= (*Z*)-5-dodecen-11-olide), isolated from *Phoracantha synonyma* Newman [113], contained two steps similar to those of **VIII/191**, see Chapter VIII.1.

---

11) Dihydrorecifeiolide was isolated from *Cryptolestes ferrugineus* [111].

The bicyclic diol, **VIII/194**, prepared according to Scheme VIII/35, was oxidized with lead tetraacetate to the ten-membered lactone, **VIII/195**. The latter was then transformed into phoracantholide I (see Chapter VII) [114].

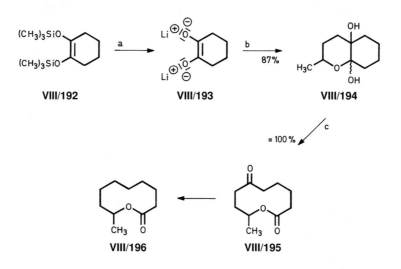

VIII/192   **VIII/193**   **VIII/194**

VIII/196   **VIII/195**

Scheme VIII/35. Part of a phoracantholide I synthesis [114].

a) CH$_3$Li   b) H$_3$CCH(OH)(CH$_2$)$_2$I, THF, hexamethylphosphoramide
c) Pb(OAc)$_4$, benzene.

Lactam formation from an oxidative cleavage of the bicyclic zero bridge is well known, for example, **VIII/197** → **VIII/200** [115]. Because of the relative positions of a ketone and a secondary lactam in the ten-membered ring compound, **VIII/199**, however the only stable structure is that of the fused five/seven ring system in **VIII/200** [116]. The substituted 3(2H)-pyrazolones **VIII/201** are opened oxidatively by periodate to **VIII/202** [117].

In an analogous manner the indole double bond in **VIII/203** was opened oxidatively (O$_2$, rose Bengal, 200 W halogen lamp, CH$_3$OH/CH$_2$Cl$_2$, 25°, 5 h) and the desired eight-membered lactam **VIII/204** was isolated in 82 % yield [118]. Compound **VIII/204** is an intermediate in a synthetic approach to the potent antitumor antibiotic mitomycin A (see Chapter VIII.2).

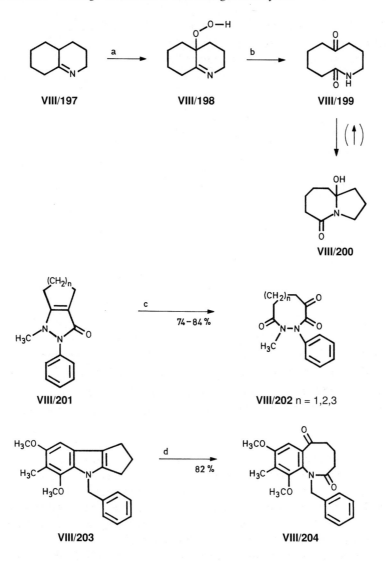

Scheme VIII/36. Lactam formation oxidative cleavage of the bridge in a bicycle.

n = 1, 2, 3
a) O_2, $AcOC_2H_5$ b) H_2O, dioxane   c) $NaIO_4$, H_2O, CH_3OH
d) O_2, $h\nu$, CH_3OH, CH_2Cl_2.

References

[1] C. A. Grob, W. Baumann, Helv.Chim.Acta **38**, 594 (1955).

[2] P. S. Wharton, J.Org.Chem. **26**, 4781 (1961).

[3] C. A. Grob, P. W. Schiess, Angew.Chem. **79**, 1 (1967), Angew.Chem.Int.Ed.Engl. **6**, 1 (1967).

[4] A. Eschenmoser, A. Frey, Helv.Chim.Acta **35**, 1660 (1952).

[5] A. C. Cope, M. M. Martin, M. A. McKervey, Quart.Rev. **20**, 119 (1966).

[6] J. March, "Advanced Organic Chemistry", 3. Ed., John Wiley & Sons, New York 1985.

[7] P. S. Wharton, G. A. Hiegel, J.Org.Chem. **30**, 3254 (1965).

[8] P. S. Wharton, Y. Sumi, R. A. Kretchmer, J.Org.Chem. **30**, 23466 (1965).

[9] J. A. Marshall, Synthesis **1971**, 229.

[10] J. A. Marshall, Rec.Chem.Progr. **30**, 2 (1969).

[11] J. A. Marshall, G. L. Bundy, J.Chem.Soc., Chem.Commun. **1967**, 854.

[12] J. A. Marshall, G. L. Bundy, J.Am.Chem.Soc. **88**, 4291 (1966).

[13] S. C. Cauwberghs, P. J. De Clercq, Tetrahedron Lett. **29**, 6501 (1988).

[14] H. A. Patel, S. Dev, Tetrahedron **37**, 1577 (1981).

[15] D. Caine, C. J. McCloskey, D. V. Derveer, J.Org.Chem. **50**, 175 (1985).

[16] M. F. Bartlett, D. F. Dickel, W. I. Taylor, J.Am.Chem.Soc. **80**, 126 (1958).

[17] C. A. Grob, H. R. Kiefer, H. Lutz, H. Wilkens, Tetrahedron Lett. **1964**, 2901.

[18] B. Witkop, J.Am.Chem.Soc. **72**, 1428 (1950).

[19] J. A. Marshall, C. J. V. Scanio, W. J. Iburg, J.Org.Chem. **32**, 3750 (1967).

[20] E. J. Corey, R. B. Mitra, H. Uda, J.Am.Chem.Soc. **85**, 362 (1963).

[21] E. J. Corey, R. B. Mitra, H. Uda, J.Am.Chem.Soc. **86**, 485 (1964).

[22] C. A. Grob, H. R. Kiefer, H. J. Lutz, H. J. Wilkens, Helv.Chim.Acta **50**, 416 (1967).

[23] M. Geisel, C. A. Grob, R. A. Wohl, Helv.Chim.Acta **52**, 2206 (1969).

[24] J. A. Marshall, J. H. Babler, J.Org.Chem. **34**, 4186 (1969).

[25] G. Read, V. M. Ruiz, J.Chem.Soc. Perkin Trans. I **1973**, 1223.

[26] M. Ikeda, K. Ohno, M. Takahashi, K.-i. Homma, T. Uchino, Y. Tamura, Heterocycles **20**, 1005 (1983).

[27] A. Fischli, Q. Branca, J. Daly, Helv.Chim.Acta **59**, 2443 (1976).

[28] G. Ohloff, J. Becker, K. H. Schulte-Elte, Helv.Chim.Acta **50**, 705 (1967).

[29] I. J. Borowitz, G. Gonis, R. Kelsey, R. Rapp, G. J. Williams, J.Org.Chem. **31**, 3032 (1966).

[30] I. J. Borowitz, G. J. Williams, L. Gross, R. Rapp, J.Org.Chem. **33**, 2013 (1968).

[31] B. Milenkov, A. Guggisberg, M. Hesse, Helv.Chim.Acta **70**, 760 (1987).

[32] B. Milenkov, A. Guggisberg, M. Hesse, Tetrahedron Lett. **28**, 315 (1987).

[33] D. Sternbach, M. Shibuya, F. Jaisli, M. Bonetti, A. Eschenmoser, Angew.Chem. **91**, 670 (1979), Angew.Chem.Int. Ed.Engl. **18**, 634 (1979).

[34] M. Shibuya, F. Jaisli, A. Eschenmoser, Angew.Chem. **91**, 672 (1979), Angew.Chem. Int.Ed.Engl. **18**, 636 (1979).

[35] G. Wilke, Angew.Chem. **69**, 397 (1957).

[36] G. Wilke, Angew.Chem. **75**, 10 (1963).

[37] A. Eschenmoser, D. Felix, G. Ohloff, Helv.Chim.Acta **50**, 708 (1967).

[38] D. Felix, J. Schreiber, G. Ohloff, A. Eschenmoser, Helv.Chim.Acta **54**, 2896 (1971).

[39] J. Schreiber, D. Felix, A. Eschenmoser, M. Winter, F. Gautschi, K. H. Schulte-Elte, E. Sundt, G. Ohloff, J. Kalvoda, H. Kaufmann, P. Wieland, G. Anner, Helv.Chim.Acta **50**, 2101 (1967).

[40] M. Tanabe, D. F. Crowe, R. L. Dehn, G. Detre, Tetrahedron Lett. **1967**, 3739.

[41] M. Tanabe, D. F. Crowe, R. L. Dehn, Tetrahedron Lett. **1967**, 3943.

[42] G. A. MacAlpine, J. Warkentin, Can.J.Chem. **56**, 308 (1978).

[43] C. Fehr, G. Ohloff, G. Büchi, Helv.Chim.Acta **62**, 2655 (1979).

[44] P. Wieland, H. Kaufmann, A. Eschenmoser, Helv.Chim.Acta **50**, 2108 (1967).

[45] D. Felix, J. Schreiber, K. Piers, U. Horn, A. Eschenmoser, Helv.Chim.Acta **51**, 1461 (1968).

[46] R. K. Müller, D. Felix, J. Schreiber, A. Eschenmoser, Helv.Chim.Acta **53**, 1479 (1970).

[47] M. Karpf, A. S. Dreiding, Helv.Chim.Acta **59**, 1226 (1976).

[48] K. Biemann, G. Büchi, B. H. Walker, J.Am.Chem.Soc. **79**, 5558 (1957).

[49] G. Rosini, J.Org.Chem. **39**, 3504 (1974).

[50] J. Jiricny, D. M. Orere, C. B. Reese, Synthesis **1978**, 919.

[51] G. L. Lange, T.-W. Hall, J.Org.Chem. **39**, 3819 (1974).

[52] M. Hanack, A. Heumann, Tetrahedron Lett. **1969**, 5117.

[53] C. E. Harding, M. Hanack, Tetrahedron Lett. **1971**, 1253.

[54] P. Borrevang, J. Hjort, R.T. Rapala, R. Edie, Tetrahedron Lett. **1968**, 4905.

[55] M. Hesse, W.v.Philipsborn, D. Schumann, G. Spiteller, M. Spiteller-Friedmann, W. I. Taylor, H. Schmid, P. Karrer, Helv.Chim.Acta **47**, 878 (1964).

[56] D. Schumann, H. Schmid, Helv.Chim.Acta **46**, 1996 (1963).

[57] M. Hesse, "Indolalkaloide in Tabellen", Springer-Verlag, Berlin 1964, Ergänzungswerk 1968.

[58] M. Hesse, "Alkaloid Chemistry", J. Wiley & Sons, New York 1981.

[59] N. J. Leonard, Rec.Chem.Progr. **17**, 243 (1956).

[60] M. P. Wentland, Tetrahedron Lett. **30**, 1477 (1989).

[61] P. Aeberli, W. Houlihan, J.Heterocycl.Chem. **15**, 1141 (1978).

[62] F. Stierli, R. Prewo, J. H. Bieri, H. Heimgartner, Helv.Chim.Acta **66**, 1366 (1983).

[63] H. Heimgartner, Wiss.Z.Karl-Marx-Univ. Leipzig, Math.-Naturwiss. R. **32**, 365 (1983).

[64] S. M. Ametamey, R. Hollenstein, H. Heimgartner, Helv.Chim.Acta **71**, 521 (1988).

[65] H. H. Wasserman, R. P. Robinson, H. Matsuyama, Tetrahedron Lett. **21**, 3493 (1980).

[66] H. H. Wasserman, M. R. Leadbetter, Tetrahedron Lett. **26**, 2241 (1985).

[67] H. H. Wasserman, R. P. Robinson, Tetrahedron Lett. **24**, 3669 (1983).

[68] H. H. Wasserman, R. P. Robinson, C. G. Carter, J.Am.Chem.Soc. **105**, 1697 (1983).

[69] H. H. Wasserman, R. K. Brunner, J. D. Buynak, C. G. Carter, T. Oku, R. P. Robinson, J.Am.Chem.Soc. **107**, 519 (1985).

[70] T. Kametani, K. Takahashi, M. Ihara, K. Fukumoto, J.Chem.Soc. Perkin Trans. I **1979**, 1847.

[71] T. Kametani, K. Takahashi, M. Ihara, K. Fukumoto, J.Chem.Soc., Perkin Trans. I **1978**, 662.

[72] H. Stetter, H. Spangenberg, Chem.Ber. **91**, 1982 (1958).

[73] D. S. Kemp, M. D. Sidell, T. J. Shortridge, J.Org.Chem. **44**, 4473 (1979).

[74] E. L. Buhle, A. M. Moore, F.Y. Wiselogle, J.Am.Chem.Soc. **65**, 29 (1943).

[75] H. Stetter, K. Findeisen, Chem.Ber. **98**, 3228 (1965).

[76] Q. Meng, M. Hesse, Synlett **1990**, 148.

[77] M.v.Strandtmann, C. Puchalski, J. Shavel, J.Org.Chem. **33**, 4015 (1968).

[78] M. Umehara, T. Oda, Y. Ikebe, S.Hishida, Bull.Chem. Soc.Jpn. **49**, 1075 (1976).

[79] T. Ohnuma, Y. Sekine, Y. Ban, Tetrahedron Lett. **1979**, 2533.

[80] T. Ohnuma, Y. Sekine, Y. Ban, Tetrahedron Lett. **1979**, 2537.

[81] W. Hückel, L. Schnitzspahn, Liebigs Ann.Chem. **505**, 274 (1933).

[82] K. Schenker, V. Prelog, Helv.Chim.Acta **36**, 896 (1953).

[83] M.v.Strandtmann, C. Puchalski, J. Shavel, J.Org.Chem. **33**, 4010 (1968).

[84] T. Fujiwara, T. Ohsaka, T. Inoue, T. Takeda, Tetrahedron Lett. **29**, 6283 (1988).

[85] L. A. Paquette, W. E. Fristad, D. S. Dime, T. R. Bailey, J.Org.Chem. **45**, 3017 (1980).

[85a] P. F. King, L. A. Paquette, Synthesis **1977**, 279.

[86] E. Vogel, O. Roos, K.-H. Disch, Liebigs Ann.Chem. **653**, 55 (1962).

[87] R. Huisgen, G. Boche, A. Dahmen, W. Hechtl, Tetrahedron Lett. **1968**, 5215.

[88] S.W. Staley, T. J. Henry, J.Am.Chem.Soc. **93**, 1292 (1971).

[89] S.W. Staley, T. J. Henry, J.Am.Chem.Soc. **92**, 7612 (1970).

[90] A. C. Cope, A. C. Haven, F. L. Ramp, E. R. Trumbull, J.Am.Chem.Soc. **74**, 4867 (1952).

[91] E. Vogel, Angew.Chem. **74**, 829 (1962).

[92] E. Vogel, Angew.Chem. **72**, 4 (1960).

[93] E. J. Corey, A. G. Hortmann, J.Am.Chem.Soc. **87**, 5736 (1965).

[94] Y. Fujimoto, T. Shimizu, T. Tatsuno, Tetrahedron Lett. **1976**, 2041.

[95] W. E. Fristad, D. S. Dime, T. R. Bailey, L. A. Paquette, Tetrahedron Lett. **22**, 1999 (1979).

[96] M. Baumann, W. Hoffmann, N. Müller, Tetrahedron Lett. **1976**, 3585.

[97] I. J. Borowitz, G. J. Williams, L. Gross, H. Beller, D. Kurland, N. Suciu, V. Bandurco, R. D. G. Rigby, J.Org.Chem. **37**, 581 (1972).

[98] I. J. Borowitz, V. Bandurco, M. Heyman, R. D. G. Rigby, S.-N. Ueng, J.Org.Chem. **38**, 1234 (1973).

[99] J. R. Mahajan, H. C. Araujo, Synthesis **1975**, 54.

[100] J. R. Mahajan, G. A. L. Ferreira, H. C. Araujo, J.Chem.Soc., Chem.Commun. **1972**, 1078.

[101] J. R. Mahajan, H. C. Araujo, Synthesis **1976**, 111.

[102] J. R. Mahajan, H. C. de Araújo, Synthesis **1981**, 49.

[103] H. Immer, J. F. Bagli, J.Org.Chem. **33**, 2457 (1968).

[104] J. F. Bagli, H. Immer, Can.J.Chem. **46**, 3115 (1968).

[105] J. R. Mahajan, G. A. L. Ferreira, H. C. Araujo, B. J. Nunes, Synthesis **1976**, 112.

[106] A. P. Krapcho, Synthesis **1982**, 805, 893.

[107] D. H. Miles, B.-S. Huang, J.Org.Chem. **41**, 208 (1976).

[108] B. M. Trost, T. R. Verhoeven, J.Am.Chem.Soc. **102**, 4743 (1980).

[109] W. S. Johnson, C. A. Harbert, B. E. Ratcliffe, R. D. Stipanovic, J.Am.Chem.Soc. **98**, 6188 (1976).

[110] B. Milenkov, M. Hesse, Helv.Chim.Acta **69**, 1323 (1986).

[111] J. W. Wong, V. Verigin, A. C. Oehlschlager, J. H. Borden, H. D. Pierce, A. M. Pierce, L. Chong, J.Chem.Ecol. **9**, 451 (1983).

[112] F. Camps, J. Coll, A. Messeguer, F. Pujol, Chem. Lett. **1983**, 971.

[113] B. P. Moore, W. V. Brown, Aust.J.Chem. **29**, 1365 (1976).

[114] T. Wakamatsu, K. Akasaka, Y. Ban, J.Org.Chem. **44**, 2008 (1979).

[115] L. A. Cohen, B. Witkop, J.Am.Chem.Soc. **77**, 6595 (1955).

[116] R. Wälchli, S. Bienz, M. Hesse, Helv.Chim.Acta **68**, 484 (1985).

[117] H. Weber, E. Wollenberg, Arch.Pharm. **321**, 551 (1988).

[118] T. Kametani, T. Ohsawa, M. Ihara, Heterocycles **12**, 913 (1979).

IX. Cleavage of the One-Atom-Bridge in Bicycles and Transesterification

IX.1. Cleavage of the One-Atom-Bridge in Bicycles

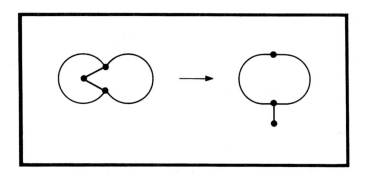

The significance of bicyclic intermediates in the synthesis of ring enlargement products was demonstrated in Chapter VIII. The bicyclic compounds being discussed there contained always a zero bridge. In this chapter we will show how bicyclic compounds with a one-atom-bridge can be cleaved in order to obtain an expanded ring. The size of the new ring is one atom smaller than the total number of ring atoms in the bicycle. The atom incorporated in the "one-atom-bridge" can be carbon, sulfur, nitrogen, and even oxygen.

The carbon bridge is normally a ketone bridge, prepared by Michael addition of an α,β-unsaturated ketone, aldehyde or corresponding substances to a cycloalkanone. After the Michael addition has taken place as illustrated in Scheme IX/1, an aldol reaction occurred because of the two free α-positions to the cyclic ketone.

The pyrrolidine enamine of cyclohexanone (**IX/1**) treated with acrylaldehyde yields the bicyclic compound, **IX/2**, in 72 % yield in which the pyrrolidine ring has moved. On heating with aqueous base, the methiodide **IX/3** was transformed to 4-cyclooctene-carboxylic acid (**IX/4**) [1]. In a similar reaction, but without reorganisation of the substituents, 2-nitrocyclohexanone (**IX/5**) was converted to the bicyclic ketoalcohol, **IX/6**, and oxidized to the diketone **IX/7**. Under very mild reaction conditions, the ketone bridge in **IX/7** is cleaved to give a quantitative yield of aldol condensation product **IX/8**. Presumably, **IX/8** was formed *via* 1,4-cyclooctanedione or 4-oxocyclooctanenitronate [2].

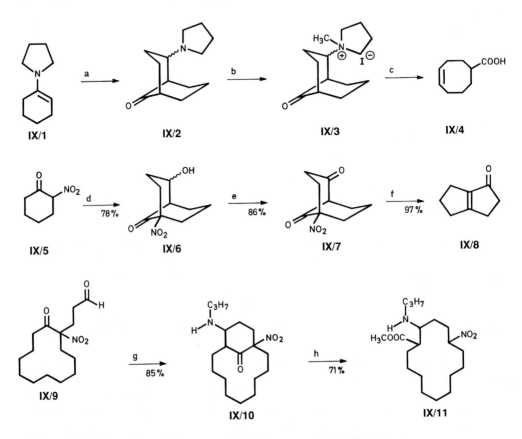

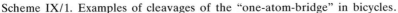

Scheme IX/1. Examples of cleavages of the "one-atom-bridge" in bicycles.

a) $CH_2=CH-CHO$, dioxane b) CH_3I c) $NaOH/H_2O$, heat
d) $CH_2=CH-CHO$, Bu_4NF e) CrO_3 f) K_2CO_3, $H_2O - H_2SO_4$
g) $H_2NC_3H_7$, THF h) CH_3ONa, CH_3OH.

The transformation of cyclododecanone *via* **IX/9** to the bicyclic intermediate, **IX/10** is possible through an internal enamine reaction. Cleavage of the central ketone bridge gives the 14-membered product **IX/11** [3]. This reaction was a key step in the synthesis of (±)-muscone (**IX/15**), Scheme IX/2, [4]. On treatment with base, the bicyclic intermediate, **IX/13**, prepared from 2-nitrocyclotridecanone (**IX/12**), was quantitatively (R=H) [5] (or in 47 % yield (R=CH₃) [4]) converted into the enlarged product **IX/14**. The retro aldol reaction was not

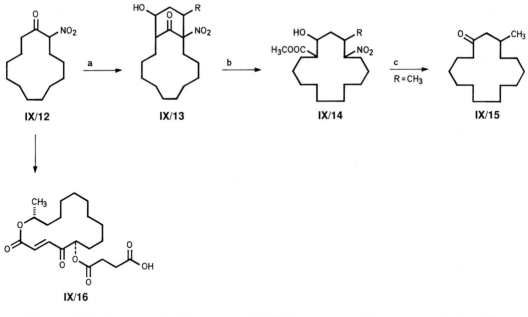

IX/12   IX/13   IX/14   IX/15

IX/16

Scheme IX/2. Synthesis of (±)-muscone (**IX/15**) by cleavage of the one-atom-bridge [4].

a) H_3C-CH=CH-CHO, Bu_3P b) CH_3ONa, CH_3OH
c) 1. CrO_3 2. Bu_3SnH, 2,2'-azabisisobutyronitrile   3. KOH, H_2O.

observed. In the synthesis of the lactone antibiotic A 26771B (**IX/16**) [6], this type of ring enlargement was also used in a key step [5].

A remarkable ring enlargement reaction was observed when the diketone, **IX/17**, (Scheme IX/3) was kept under acetalization conditions (BF_3-etherate/ethyleneglycol) [7]. The seven-membered **IX/20**[1] was isolated in 98 % yield. Probably this reaction is restricted to five- to seven-membered ring enlargement[2].

From a mechanistic point of view, the first step is an acid catalyzed aldol reaction to **IX/18**. Acetalization of the remaining ketone, **IX/19**, and cleavage of the one-atom-bridge led to **IX/20**. This reaction was applied to the synthesis of bulnesol, **IX/22**, using the diketone, **IX/21**, as a starting material [7].

Carbon monoxide elimination is observed when the bicyclic compounds of type **IX/24** decompose [8]. Depending on the nature of the substituents at the bicyclic intermediate, **IX/24** is more or less stable. Compound **IX/24** and its dihydroderivative can be prepared by Diels-Alder reaction of a cyclopentadie-

---

1) The reaction mentioned, was the most efficient of a series.
2) A five- to eight-membered conversion (prolongation of the side chain by one CH_2-group), was not successful [7].

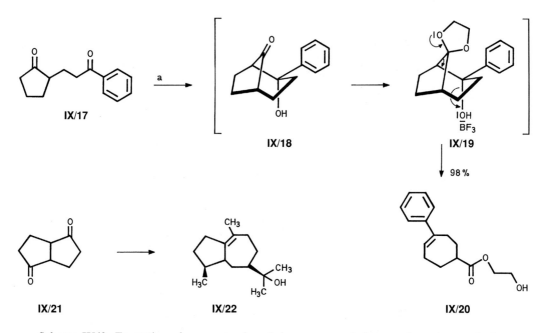

IX/17 a → **IX/18** **IX/19** ↓ 98 %

IX/21   **IX/22**   **IX/20**

Scheme IX/3. Formation of seven-membered ring compounds from cyclopentanone derivatives [7].

a) $BF_3x(C_2H_5)_2O$, $(CH_2OH)_2$.

none (*e.g.* **IX/23**) and an alkyne or an olefin. This reaction is an useful method for preparation of specifically substituted benzenes and cyclohexadienes (if a dihydroderivative of **IX/24** is heated, Scheme IX/4). Further examples are given in ref. [9].

The sulfur-mediated total synthesis of the biologically potent zygosporin E (**IX/31**) was published [10] and the important ring enlargement steps are given in Scheme IX/5. When compound **IX/26** was heated with NaI/K_2CO_3 in acetonitrile, the nine-membered **IX/28** was formed. The medium sized ring is built up by a rearrangement of the first formed six-membered sulfonium ion **IX/27** (see

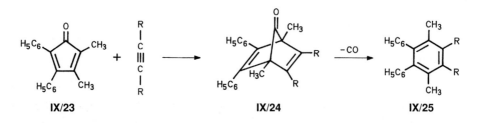

IX/23   **IX/24**   **IX/25**

Scheme IX/4. Formation of aromatic compounds by CO extrusion [8].

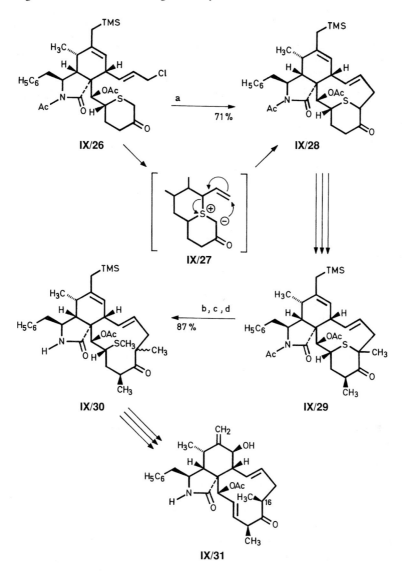

Scheme IX/5. Synthesis of zygosporin E (**IX/31**) with two sulfur-mediated ring expansions [10].

a) NaI, K_2CO_3, CH_3CN, heat b) $(CH_3)_3OBF_4$
c) Zn, 1,2-dimethoxyethane, THF, HOAc, 20° d) K_2CO_3, CH_3OH.

Chapter V) [11]. Afterward, the sulfide bridge was methylated by the Meerwein reagent, followed by Rieke zinc treatment to get the central eleven-membered ring. N-Deacetylation (K_2CO_3/CH_3OH) gave **IX/30**, which was finally transformed to (±)-zygosporin E (**IX/31**) and its 16-epimer. The synthesis of zygosporin E illustrates the use of the stereochemistry of the sulfide bridge as a relay of stereochemical information in medium-sized rings [10]. Further applications of this method are given in ref. [12] [13].

A ring expansion reaction not easily classified is represented by a reaction type in which macrocyclic lactones are formed through a sulfide contraction. The reaction principle is shown in Scheme IX/6. The first step is a ring closure reaction by formation of a sulfur carbon bond, followed by an additional carbon carbon bond formation to give an episulfide. The sulfur bridge is then removed by phosphine (sulfide contraction method [14]). The resulting compound, **IX/36**, is a β-ketolactone [15]. This reaction has been used to synthesize different medium and large ring macrolides, *e.g.* ($\pm$)-diplodialide A (**IX/37**) [15] [16] [17] [18].

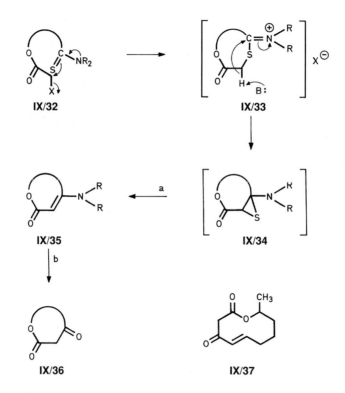

Scheme IX/6. Macrocyclic lactone formation through sulfide contraction [15].

a) R'_3P b) H_3O^+.

Raney nickel can be used to cleave a sulfur bridge in a bicycle as well as of a thiophene moiety in a macrocyclic compound. Thus, thiophene derivative **IX/38** cyclized by an intramolecular acylation to the lactone **IX/39** in good yields

(48–68 %) [19]. Reductive desulfurization converts **IX/39** to the corresponding ketolactone, **IX/40** (70–80 %) [20].

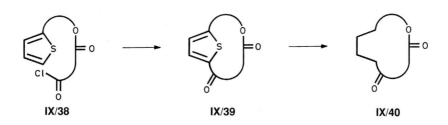

A method involving an oxygen bridge as a synthetic tool has also been published [21]; the synthesis of the aglycone of the antibiotic neomethymycin, neomethynolide (**IX/45**) [22]. The acetylenic fragment, **IX/41**, was lithiated and condensed in tetrahydrofuran with the Prelog-Djerassi lactonic acid methyl ester, **IX/42**, to give a mixture of hemiacetals. These were then converted to the methylacetals, **IX/43**. By a ring closure reaction (Scheme IX/7), the bicyclic,

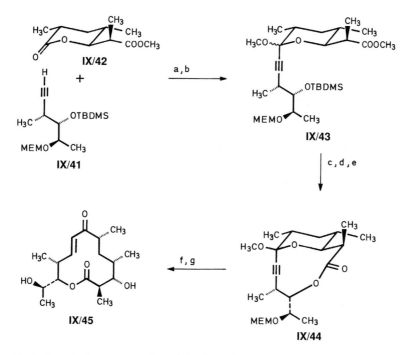

Scheme IX/7. Synthesis of neomethynolide (**IX/45**) [21].

a) BuLi, THF   b) TsOH, CH₃OH   c) Bu₄NF, THF   d) NaOH, CH₃OH
e) 2,4,6-trichlorobenzoyl chloride   f) ZnBr₂, CH₂Cl₂, CH₃NO₂
g) CrSO₄, dimethylformamide, H₂O.

highly strained lactone **IX/44** was obtained. The oxygen bridge in **IX/44** is part of an acetal, and, when cleaved, gave the twelve-membered ring in **IX/45**. The (*E*)-double bond in **IX/45** was generated in aqueous dimethylformamide by reduction with chromous sulfate [23]. For similar reactions see ref. [24].

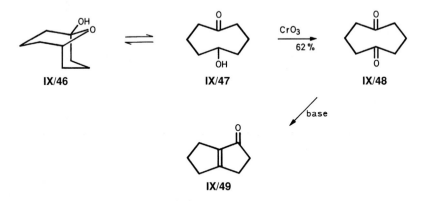

Scheme IX/8.

5-Hydroxycyclooctanone (**IX/47**) exists almost completely in the transannular hemiacetal form, **IX/46**, both as a solid (the infrared spectrum in potassium bromide shows no carbonyl absorption), and in solution (carbon tetrachloride). Oxidation of **IX/46** to the diketone, **IX/48**, with chromic anhydride proceeds in good yield because an aqueous solution contains about 4 % of **IX/47** [25]. The diketone **IX/48** is readily cyclized by a base catalyzed transannular aldol reaction [2] to give **IX/49**. The synthesis of cyclooctene and cyclononene derivatives was realized using bridged-ring precursors [26] [27]. Reactions involving a one-atom-bridge containing nitrogen were mentioned in ref. [28].

IX.2. Transesterification

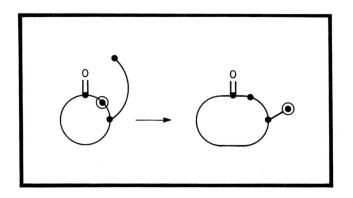

Comparison of transesterification with transamidation (Chapter VI.1) is of interest. The main difference is that oxygen forms two bonds and nitrogen forms three bonds. The driving force in large ring transamidation reactions is the formation of a resonance stabilized secondary amide. If esters or lactones are involved, no such resonance stabilizing structures are possible. Therefore, in lactones, effects of lesser importance may influence an equilibrium mixture of ring contraction and ring enlargement products. These effects, which are not specific for oxygen containing compounds, are known as steric release effects. They will be observed when a medium-sized ring becomes larger or if a sterically crowded ring can be released by enlargement.

The formation of the twelve-membered lactone, **IX/52**, from the eight-membered ketone, **IX/51**, prepared by alkylation of the aldehyde, **IX/50**, can be explained as an enlargement of a medium-sized ring. The driving force is probably the resonance stabilisation of the secondary nitro group in base (nitronate), compared to the tertiary nitro group in the starting material (Scheme IX/9) [29].

The second step, transformation of the twelve-membered **IX/53** to the 14-membered lactone **IX/54**, gives an equilibrium mixture of **IX/53** and **IX/54** (ratio approx. 1:2). The ratio is clearly different from a 1:1 mixture, which might be expected if both compounds contained the same functional groups and rings and were strained. Other reactions are analogous (Scheme IX/10): The strained nine-membered lactones **IX/55** and **IX/57** are converted to the twelve-membered **IX/56**, and, *via* a double translactonization, to the 15-membered **IX/58** (a mixture of *cis* and *trans* isomers), respectively. In both cases, the reactions proceed better under acidic than under basic conditions. While the seven-membered **IX/59** does not isomerize to the more strained ten-membered **IX/60**, its methyl derivative, **IX/61**, is converted into an equilibrium mixture of **IX/61** and

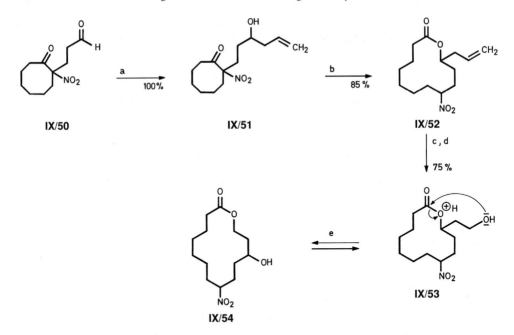

IX/50   **IX/51**   **IX/52**

IX/54   **IX/53**

Scheme IX/9. Transesterification [29].

a) CH$_2$=CH-CH$_2$-Si(CH$_3$)$_3$, TiCl$_4$, CH$_2$Cl$_2$ b) Bu$_4$NF, THF
c) O$_3$, CH$_3$OH   d) NaBH$_4$ e) camphor-10-sulfonic acid, CH$_2$Cl$_2$.

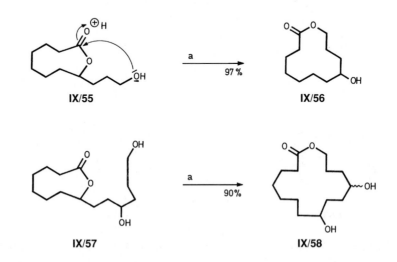

IX/55   **IX/56**

IX/57   **IX/58**

Scheme IX/10. Examples of transesterification [30].

a) TsOH, CH$_2$Cl$_2$, 25°, 2 h.

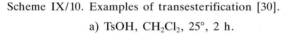

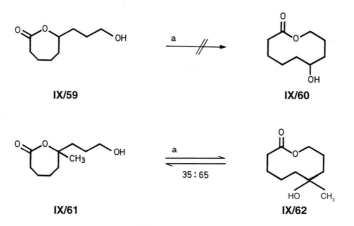

Scheme IX/10 (continued).

IX/62 by storage at room temperature (pure or in chloroform solution) for three days. As in the case of transamidation reactions, eight-membered intermediates are much less favored than seven-membered rings. Thus, translactonization will not take place.

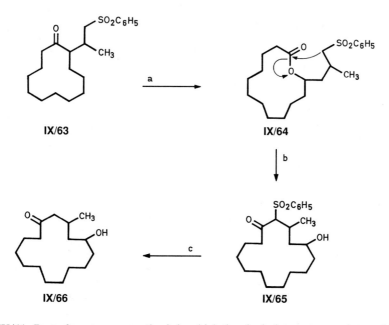

Scheme IX/11. Part of a muscone synthesis in which the alcoholate acts as an internal leaving group [31].

  a) CH₃CO₃H, BF₃, (C₂H₅)₂O, Cl₃CCH₃
  b) lithiumdiisopropylamide, THF   c) 1. Al/Hg 2. THF, H₂O.

In the translactonization process the alcoholate acts as an internal leaving group from displacement by a side-chain alcohol. However, nucleophiles other than an alcohol may also displace. This can be a carbon atom in an α-position to a sulfone group as in **IX/64** (Scheme IX/11) (various bases were used). The 15-membered **IX/64** is isolated in 80 % yield. This synthesis represents a transformation of cyclododecanone *via* **IX/63** to **IX/66** to muscone [31].

Instead of the carbanion, stabilized by the sulfone group, a Grignard reagent can be prepared from a side chain bromide. This reacts with the carbonyl group of the lactone in a similar way. The transformation (not transesterification) of **IX/67** → **IX/68** is only realized in a yield of 37 %, Scheme IX/12 [31]. Lactone to carbocycle transformations are rare; the method presented is a possibility for the realization of such a conversion.

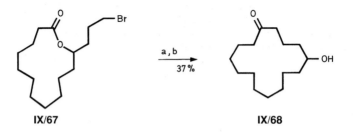

IX/67   IX/68

Scheme IX/12. a) Mg, THF   b) 1. $BF_3 \cdot (C_2H_5)_2O$, 2. NH_4Cl, H_2O.

Translactonization reactions (thiolactone → lactone) are discussed in Chapter V, beginning with Scheme V/21.

References

[1] G. Stork, H. K. Landesman, J.Am.Chem.Soc. **78**, 5129 (1956).
[2] A. Lorenzi-Riatsch, Y. Nakashita, M. Hesse, Helv.Chim.Acta **67**, 249 (1984).
[3] A. Lorenzi-Riatsch, R. Wälchli, M. Hesse, Helv.Chim.Acta **68**, 2177 (1985).
[4] S. Bienz, M. Hesse, Helv.Chim.Acta **70**, 2146 (1987).
[5] S. Bienz, M. Hesse, Helv.Chim.Acta **70**, 1333 (1987).
[6] K. Tatsuta, A. Nakagawa, S. Maniwa, M. Kinoshita, Tetrahedron Lett. **21**, 1479 (1980).
[7] M. Tanaka, H. Suemune, K. Sakai, Tetrahedron Lett. **29**, 1733 (1988).
[8] C. F. H. Allen, Chem. Rev. **62**, 653 (1962).
[9] M. A. Ogliaruso, M. G. Romanelli, E. I. Becker, Chem.Rev. **65**, 261 (1965).
[10] E. Vedejs, J. D. Rodgers, S. J. Wittenberger, J.Am.Chem.Soc. **110**, 4822 (1988).
[11] E. Vedejs, J. G. Reid, J.Am.Chem.Soc. **106**, 4617 (1984).
[12] E. Vedejs, R. A. Buchanan, P. C. Conrad, G. P. Meier, M. J. Mullins, J. G. Schaffhausen, C. E. Schwartz, J.Am.Chem.Soc. **111**, 8421 (1989).

[13] E. Vedejs, R.A.Buchanan, Y. Watanabe, J.Am.Chem.Soc. **111**, 8430 (1989).

[14] M. Roth, P. Dubs, E. Götschi, A. Eschenmoser, Helv.Chim.Acta **54**, 710 (1971).

[15] R. E. Ireland, F. R. Brown, J.Org.Chem. **45**, 1868 (1980).

[16] T. Ishida, K. Wada, J.Chem.Soc., Chem.Commun. **1977**, 337.

[17] T. Ishida, K. Wada, J.Chem.Soc. **1979**, 323.

[18] T. Ishida, K. Wada, J.Chem.Soc., Chem.Commun. **1975**, 209.

[19] Y. L. Goldfarb, S. Z. Taits, F. D. Alashev, A. A. Dudinov, O. S. Chizhov, Khim.Getero-sikl.Soedin. **1**, 40 (1975).

[20] S. Z. Taits, F. D. Alashev, Y. L. Goldfarb, Izv.Akad.Nauk S.S.S.R., Ser.Khim. **3**, 566 (1968).

[21] J. Inaga, Y. Kawanami, M. Yamaguchi, Chem.Lett. **1981**, 1415.

[22] C. Djerassi, O. Halpern, Tetrahedron **3**, 255 (1958).

[23] C. E. Castro, R. D. Stephens, J.Am.Chem.Soc. **86**, 4358 (1964).

[24] J. Inanaga, T. Katsuki, S. Takimoto, S. Ouchida, Chem.Lett. **1979**, 1021.

[25] G. I. Glover, R. B. Smith, H. Rapoport, J.Am.Chem.Soc. **87**, 2003 (1965).

[26] G. L. Buchanan, A. McKillop, R. A. Raphael, J.Chem.Soc. **1965**, 833.

[27] H. Marschall, F. Vogel, Chem.Ber. **107**, 2176 (1974).

[28] S. Sakane, K. Maruoka, H. Yamamoto, Tetrahedron Lett. **24**, 943 (1983).

[29] T. Aono, M. Hesse, Helv.Chim.Acta **67**, 1448 (1984).

[30] E. J. Corey, D. J. Brunelle, K. C. Nicolaou, J.Am.Chem.Soc. **99**, 7359 (1977).

[31] C. Fehr, Helv.Chim.Acta **66**, 2512 (1983).

Compound Index

The formation of a ring-enlarged product is indicated by an arrow preceding a compound's name. An arrow at the end denotes that the corresponding compound is used as starting material in a ring-enlargement reaction.

Subject Index

A

Abelmoschus moschatus   1
Acetaldehyde, (phenylseleno)-   80
Acetate, butenyl   149
Acetate, diazo-   3, 11
Acetate, phenyl-   57
Acetic formic anhydride   33
Acetophenones   65
Acetyl chloride, phenyl-   57
Acetylene carboxylate   59, 61-63
Acetylene dicarboxylate   59, 60, 62-63
Acoragermacrone   77
Acridone   32
Actomyces   1
2-Adamantanone   9
Aldol reaction   49, 77, 163, 200, 201, 206
Alkanes, diazo-   9
Alkanimide, ω-phenyl-   68
Alpine-hydrides   147, 148
Aluminium chloride, diethyl   14, 15
Ambrette seeds   1
Androstane derivatives   136, 137
Angelica roots   1
Anthracene, perhydro-   166
Anthraquinones   55
Antibiotic A 26771 B   32, 201
Archangelica officinalis   1
Arynes   67
Arynic condensation   **67-69**
2-Azabicyclo[4.4.0]decenes   195
1-Azabicyclo[11.3.1]heptadecane   144
2-Azabicyclo[3.2.0]heptane   65
7-Azabicyclo[4.1.0]heptane   43
1-Azabicyclo[3.2.0]heptenes   25, 181
16-Azabicyclo[10.3.1]hexadecene   25
13-Azabicyclo[10.3.0]hexatriene   23
16-Azabicyclo[10.3.1]hexatriene   23
2-Azabicyclo[3.2.2]nonanone   25
7-Azabicyclo[4.3.0]nonene   65

3-Azabicyclo[3.2.1]octanes   30
3-Azabicyclo[3.2.1]octene   30
2-Azabicyclo[3.3.0]octenone   31
3-Azacyclotetradecane   144
7-Aza-10-decanelactams   100
4-Aza-7-heptanelactams   107
13-Aza-16-hexadecanelactams   103, 104
17-Aza-20-icosanelactam   110
4-Aza-8-octanelactams   108
1-Azaspiro[4.11]hexadecene   130
8-Aza-11-undecanelactam   99, 100
Azepines   28, 44,
Azepinones   29, 86, 181
Azetidinium salt   85
Azetidinones   11, 23, 27, 39, 40, 44, 86, 99, 105-108, 111-113, 115, 117, 118
Aziridines   **39-45**
Azirines   39-42, 120, 121
Azocines   85
3*H*-Azonines   85

B

Baeyer-Villiger rearrangement
   (*B.-V.*oxidation) 3, **32-34**, 190
Barton reaction   136, 137
Beckmann rearrangement   20, **24-32**, 103
Benzamides   28, 102
Benzazepines   59, 61, 85
Benzazocines   61, 85, 181, 195
Benzazonines   59, 84
Benzisothiazole   41
Benzocycloalkenones   55, 67-69, 81, 127, 168
Benzoate   28
Benzocycloalkanes   17
Benzocycloalkenes   59, 67, 81
Benzocyclobutenones   55
Benzocyclodecenones   81
Benzocyclododecenones   69